Collecting
Transistor
Novelty Radios

Robert F. Breed

ISBN 0-89145-446-2

L-W BOOK SALES
P. O. Box 69
Gas City, IN 46933

Acknowledgments

Few books on collecting are written alone. This one is no exception. Many people have helped and encouraged me on this journey.

First, I would like to thank Lee and Carolyn Bruton (LCB) of Longmont, Colorado. Lee not only provided encouragement, he agreed to photograph his collection of novelty sets using the same backdrop as mine — keeping the book somewhat uniform. A heartfelt thanks for his time and effort.

Several collectors in the San Diego area also made their collections available for me to photograph: Bob and Nancy Zelenack (BNZ), Walt and Mary Curry (WMC), and Bill and Susan Gabb (BSG) were most patient while I posed, photographed, and measured each unit in their collections. Their patience and other extended courtesies are appreciated.

Walt also provided most of the background information on the Regency TR-1 radio, and provided editorial comments on the rest of this book. A final polish was placed on the text by Thomas K. Arnold of the *Los Angeles Times*. Their efforts have made the book much more concise — and minimized my ramblings.

A special thanks to Jerry Koening, of Premium Resources International, who contributed a wealth of information about product-shaped radios. Much of this information appears in the chapter on advertising and product-shaped radios.

The warmest thanks go to my wife and helpmate, Jean, who was always by my side during the hunt for the radios that appear in my collection, and for her support during the long hours to photograph, caption, and compile this book.

v

Table Of Contents

Introduction vii

A Look Back 1

About The Pictures 5

Travel & Transportation 7

Media Stars & Heros 33

Character & Figural 69

Robots, Rockets & Outer Space 91

Advertising & Product Shaped 101

Food & Drink 125

Musical & Home Entertainment 151

Sports & Recreation 167

Household & Personal Use 177

Warfare, Warriors, & Weapons 191

Miscellaneous 197

Building A Collection 205

Price Guide 209

Bibliography 217

Introduction

Collectors are often asked, "Why did you start collecting **THOSE** things?" While "those things" might be toys, dolls, beer cans, or just about any other item you may mention, I'm asked this question because my collection - and this book - consists of radios that are different than the ones most radio collectors seek: namely, the vintage sets of the 1920s through the 1940s. While my fellow radio collectors think I've lost my mind, the answer to their question is really very simple: I like them! I believe the time to build a collection of transistor novelty units is NOW, while they're available, inexpensive, and still represent a new field for collectors. I hope others will share my enthusiasm.

I have a large collection of the aforementioned vintage sets and have watched the prices of those radios increase steadily. That generation of radios is no longer affordable, nor are they easy to locate. If you were to try to build a collection of vintage radios today it would require a rather large bankroll - and lots of patience!

Still, I can hear you scoff, "I think of transistor novelty radios as cheap little toys - and on top of that, they're made of PLASTIC (a word that's almost obscene to a **TRUE** vintage radio collector - and even more contemptible to an 'olde' toy buff)." True, most of the radios in this book *are* of this scorned material, but many of them are stamped (or cast) metal. A precious few are even cased in wood. As for the "cheap toy" designation, a quick look through this book might change your mind.

In truth, plastic is not a new product to the radio fraternity. It first appeared in the early 1930s in several tabletop radios, usually in "Art Deco" or other free-form designs. These radios were made of Catalin, Celluloid, Beetle, or the lowest of all the plastics, brown Bakelite (often painted other colors).

During the period while I was concentrating on collecting only wooden-cased radios, many fine Catalin and other "Art Deco" types in plastic cases were passed over. These were often less than $5 "as is." Today, a very active group of collectors pursue these very same radios - and they command a higher price than the wooden-cased ones I so ardently sought! (Many of the Catalin types are in the $500 range, and a 1930s blue-glass Sparton sells for over $2000.)

Will the novelty radios pictured in this book ever command those prices? Probably not, but they will be valuable for the same reason anything old is collectible: The nostalgia factor! When the children of today reach middle age, some of them will become collectors. Collectors are often motivated by those things that trigger fond memories of their youth. Many novelty radios today, especially the character ones, are purchased by the parents for the child. As the child grows older, perhaps he or she will see one of these radios at a swap meet or an antique mall, and they will flash back and remember, "Gee . . . My folks gave me one of those for Christmas back in the 1980s." A warm glow of nostalgia will sweep over them and another collector will join the ranks.

These radios may be plastic, but they will trigger the responses as well as any other item made, just as a wooden-cased radio triggers memories of my youth and the days when I would rush home from school and listen to Jack Armstrong and Little Orphan Annie, then end my day by sharing an adventure with the Lone Ranger.

Even without the nostalgia, the radios are interesting in their own right. Most people can appreciate a well-made and cleverly designed item. Some of these radios display careful thought in concealing the controls or integrating the controls into normal functions of the unit. Many of the items selected for radios are unique. I could see making a radio out of a model car - but one representing an outboard motor, a balance scale, or a ship's telegraph? That's real imagination! This type of design and craftsmanship should be preserved for future generations.

Collect them to ENJOY! If they do appreciate in value, so much the better. At this point in the hobby, these radios are still comparatively inexpensive. A large collection of them has as much appeal and variation as any collection of beer cans or Beam bottles, and they're much more functional (unless the cans and bottles happen to be full!).

I have my collection displayed in a large glass-fronted case. This collection has been shown to every age group from teenagers to grandparents, and regardless of their age, the first statement out of their mouths is inevitability, "Is everything in that case really a radio?" Upon assuring that indeed they are, the next statement (as they continue their scan) will be "Now THAT can't be a radio — I just don't believe it!" At this point, the contested unit is pulled from the case and I point out the knobs, speaker etc. The final statement is now made: "That is just darling (precious, unbelievable, etc.) - where do you find them?" While the object in dispute may vary, the final statement will always appear. Regardless of the age group, something in that case will appeal to the viewer. (And my grandkids go crazy when they see all of the character units, wondering why "mean old grandpa" won't let them play games with them. A term that I must have in common with any vintage toy collector.)

For those who don't have a collecting hobby, I would like to welcome you to the fold. Look at the pictures, read the section on building a collection, and begin a new adventure. Good hunting!

A LOOK BACK

Webster's New World Dictionary defines "Novelty" as:
      1. the quality of being novel; newness. 2. something new, fresh,
or unusual. 3. *usually in pl.* a small, often cheap, cleverly made
article.

While these classic definitions have all applied to the art of radio design since its inception, the novelty radio collector would add a fourth definition, perhaps:

      4. Anything that looks like something else, but contains a working radio.

Or

      4. A radio made out of anything not normally recognized as being a radio

If the designer has integrated normal functions of the above item into the tuning and volume controls, the radio is much more collectible than one where a small printed circuit board has been fitted into it with the controls jutting out and detracting from the item's original appearance.

While this book concentrates on the transistor novelty radio, we must go back into radio's beginnings to describe the evolution of the novelty set.

The concept of novelty radio starts almost with radio itself. In the 1920s, a crystal set called "Monte Blue" featured a man sitting in an over-stuffed chair. This may be the earliest-known novelty radio.

During the 1920s, little effort was spent in packaging the radio. This period was the era of the battery operated units, using heavy storage batteries, long wire antennas, and sturdy grounding systems. Radio itself was the novelty, and this period was more involved with technology changes than cabinet design.

From 1930 to 1960 many different novelty sets were made using vacuum tubes. The early sets in this group, such as the Mickey Mouse and Peter Pan units of the early '30s had to overcome many problems associated with heat and the sheer size of the tubes and other components in use during that period - a tribute to the designers. Other sets from this era include the beer bottle, the Colonial globe, and a beer-keg unit.

By 1935, the octal based tube started to be used. This tube, and other design improvements - mainly a reliable AC-DC circuit which removed the bulky power transformer from the chassis - allowed an even smaller radio to be developed. Radios using these tubes are the Snow White, whiskey bottles, ship radios and many of the Catalin and other "Art-Deco" radios so collectible today.

The postwar years introduced a small 7 pin tube without a base. This tube, called the "Peanut," and other smaller components, made possible a very compact chassis. While heat still had to be considered, size was only a marginal consideration. One very active company, Guild Radio, made several interesting sets using these tubes: A carriage lamp, a wall telephone, a hurdy-gurdy, and even an old-style gramophone. Other companies made novelty sets in the shape of a microphone, a remake of the 1930s' beer bottle, and in a statue of a horse. (There are many other examples as well.)

The days of vacuum-tube units finally drew to a close. The beginning of the end took place in July of 1948, when a team of Bell Lab engineers — John Bardeen, William Shockley, and Walter H. Brattain — announced the invention of the transistor. This device, operating on low voltage and low currents, generated little or no heat; it was one tenth the size of a miniature tube, shock resistant, and would theoretically last forever. It was the device that opened the door to our present solid state electronic wonders such as the personal computer, the VCR, and digitally controlled cameras - an invention so important that this team was granted the Nobel prize for physics in 1956.

Although Bell Labs granted other companies manufacturing licenses in the fall of 1951, the early years had little effect on the radio market. Low yields, uneven quality control, and the requirements for new design techniques limited its use to military applications, some early telephone repeaters, and a few hearing aids.

This was soon to change. For the Christmas season of 1954, Regency Electronics, using a circuit designed by Texas Instruments, introduced the world's first commercial transistor radio: The Regency Model TR-1. This "shirt pocket" radio, measuring only 3x5x1-1/4 inches, was easily one of the smallest radios made at that time. This radio sold for $49.95 - about five times the cost of an equivalent table-top tube unit of the same period. This tremendous difference was not just due to the cost of the transistors, which cost about $2 each as opposed to 60 cents for a tube, but also represented the development charges associated with making the miniature components for the small chassis - all of which had to be designed and manufac-

tured just for this radio. In spite of this price difference, this radio would see sales in excess of 100,000 units in its first year of production, and would remain in production for the next several years.

Although the transistor and the first shirt-pocket radio were American inventions, we failed to continue our advantage. The small profit margin in the TR-1, even at the $49.95 sale price, led Regency and Texas Instruments to make a decision to leave the consumer marketplace, a void quickly filled by the Japanese.

In August of 1955, a mere nine months after the introduction of the TR-1, Sony, a Japanese firm known for its high quality tape recorders, introduced Japan's first transistor radio. Although Sony's initial unit was too large to fit into a standard shirt pocket (a problem solved by making a shirt with an extra-large pocket just to fit this set), the door was opened for Japan to dominate this field. By March of 1957, Sony had solved the miniature component problem and the initial production of 500,000 Model TR-63s was completed that year - 75% of which went into the export market. The Japanese, with their cheap labor and very innovative designers, must be credited with making the transistor radio, and the companion novelty units, a force in the American market.

The Japanese had the foresight to commit themselves to developing a completely transistorized technology, electing to turn their backs on 50 years of proven vacuum-tube design techniques. They took the initiative to improve and mass produce the miniature transformers, speakers and tuning capacitors that made possible a small but efficient radio on a printed circuit board - at a very low cost! (An arena where U. S. companies couldn't compete.) With the development of this miniature PCB, size and heat were no longer a factor. Freed from these age-old bugaboos, the Japanese radio designers could let their imaginations soar - and soar they did. The early novelty units from Japan, with their clever designs and their use of metal or wooden cabinets, set new standards for novelty radio packaging.

Today, Japan's contribution to the novelty radio market is minimal. The cheap labor is in Hong Kong, Taiwan and Korea - and radios are now being made in China, Mexico, Singapore, and the Philippines. None of these countries have been able to match Japan for quality, but Hong Kong - and a few American companies - have proven to be equally innovative with their designs. (Many of the early Hong Kong units were only copies of previous Japanese types, although made in plastic instead of metal.)

This book, with more than 500 pictures, tries to represent a cross section of these radios, covering everything from advertising items to childrens' toys - from excellent design to the cheap and shoddy. However, this book should NOT be considered a definitive work - merely a representative one. I doubt if any one collector, or even random group of collectors, could produce a copy of every transistor novelty radio made - nor will I even hazard to guess how many different ones exist! At this time, no book can be definitive, as new radios are designed on a continuing basis, especially in the media and advertising markets. Those who use this book only to seek the radios NOT pictured will have an enjoyable future.

REGENCY MODEL TR-1. "Genesis Transistor." The world's first commercial radio using transistor technology. Made in USA and measures 3 x 5 x 1-1/4 inches. (See text).

NOTES

About the Pictures

Since I believe radio collecting is a hobby, not an art or a profession, this book is photographed, written and captioned in a very informal style. The pictures are not intended to be "portraits" of radios - rather only to show a clear and focused subject with enough detail to allow the collector to identify the unit. This picture, when combined with the information in the caption, allows the reader to make one important decision: Do I want to add this radio to my collection?

While writing the captions and the chapter introductions, I have tried to write as one collector talking to another. For this reason they may contain some personal anecdotes, opinions, conclusions, or other comments. I hope these add to your enjoyment of this book.

The captions are all in the same format. The first line contains a Plate number for use in referring to the various units among the collecting fraternity. It is also used to index the unit in the price guide.

Within the body of the caption I have included the name of the unit and the material used in its construction (if not mentioned, it is assumed to be plastic). It will also list distributor, patent, and copyright information, if known. Many times this information appears only on the box, or on the battery compartment cover - items almost certain to be lost if the radio was purchased in the secondary market. (Manufactures take notice!) Two or more copyrights may appear on some radios, especially the character ones - one for the character, and one for the radio. A few Hong Kong sets also show British as well as American patents. Copyright, patent, and trademark information is a very sensitive issue and every effort was made to list the holders if it was shown on the unit or could be obtained through other sources.

A word about distributors as opposed to manufactures: I frequently get phone calls from people offering to sell me a radio. Somewhere in the conversation, a statement will be made like this: "It sure is a beautiful unit, and it's made by Waco!" (Or Heritage, Amico, Stewart, etc.) These names are NOT manufacturers - they're importers and distributors. If the unit is Japanese, a clue to the manufacturer is a small three digit number that appears just below the words "Made In Japan." While many different numbers appear, the most common one in my collection is 611. Radios marked with this number are usually interesting and high-quality units, using metal and plastic construction. While one production run using this number might have been made for Waco, the next run could easily have been for another distributor. Researching the exact meaning of these numbers could be a very interesting project, but for now they have only passing interest and are not included in these captions. (Perhaps a future book will include this information?)

American companies - such as Concept 2000, Durham, and Philgee that distribute toys made in Hong Kong, Taiwan, and other countries are more complex. It is possible that these large companies actually own plants located on foreign soil, but for uniformity in captions I have listed them as distributors, not manufacturers.

The caption also includes the major dimensions and the country of manufacture. Measurements are taken at the maximum points, and are expressed as the unit is posed in the picture. Some measurements may be "eyeball" distorted, but most will be within 1/4 of an inch.

On those radios that have well-integrated or concealed controls, a statement may be included with this information. The majority of these radios use a control I call a "thumbwheel." This is a control that is approximately 1/8th of an inch thick, with one edge exposed to allow you to move it with your thumb or finger. Where these controls are used, or other types are plainly visible, I will omit any reference to them.

The final line contains the initials of the contributor of the radio. All contributors are given credit in the Acknowledgments section. If this space is blank, the radio appears in the author's collection.

A little bit of information for the photographers: All of my pictures were taken with a 35MM Minolta Maxxum 7000 equipped with a 28/135 AF lens, usually used in the macro mode. Film was Kodak VR-G 200. Most of the pictures were taken against a background of feather gray cloth, about one stop brighter than 18% gray reference. Because of computerized film processing, this background appears from sky blue to white. (And once in a while, even feather gray!) This cloth was stretched over two boards clamped in the shape of an "L," but leaving a slight bow at the junction to make a "seamless" background. The pictures were taken in natural light, but never in direct sunshine. To obtain depth of field, small lens openings (F16) were used and the camera was supported on a tripod. Some shots show other backgrounds that were taken during my experimental days. Lee used the same background and lighting methods for his shots, but his camera is a Nikon with a 50MM fixed-focal length lens.

Travel and Transportation

This chapter features some of the devices that man has used over the ages for travel or transportation — both for work and for play. Travel items have always held interest for manufactures of novelty units: Avon and Beam bottles, salt and pepper shakers, banks, and many other novelties have been based on these units.

This chapter features an interesting mix of these items: Cars, boats and airplanes are the most common, but this section also includes other items such as a ricksha, a few locomotives, and even stage coaches. Other miscellaneous items to aid the traveler, such as gas pumps to supply the fuel, or a life preserver in case of a nautical accident, are also shown.

This chapter is an ideal lead-off one, as it features a high percentage of the early metal units from Japan, allowing us to compare between these Japanese units and the later plastic copies made in Hong Kong.

Although variations of radios appear in every chapter in this book, I will use this one to illustrate two common techniques used to make them. The first is the use of one mold to make "different" units, usually by just changing paint, decals, or trim. The second method is more complex, and somewhat harder to spot, as the makers use only pieces of a previous unit and combine them with other new parts, or arrange them in a different way, to make a new radio. Once these methods are explained, spotting these variations can become an interesting part of the hobby.

Plates 1 and 2 illustrate the latter method. The Rolls Royce shown in Plate 1 is a very popular car that appears in many collections. It is quite distinctive, and both the Hong Kong and the Japanese versions use license plate number SD8451. Note the position of the spare tire, the hand brake (partially obscured by the spare tire), and the controls just above the running boards. Comparing it to Plate 2, we see the exact same body, even showing identical rivet patterns, but cast in white plastic. The spare tire has been relocated to the rear of the car (not visible in this photo) and the roof has been modified to a convertible type. The next step was to change the license plate number to 3288 — and like magic, another "new" car for the collector. This is a somewhat major modification, but allows two cars to be made by using the same basic mold. The simpler technique is illustrated by the Volkswagen units in Plates 19 and 20. Other than different paint and designations, the cars are identical, even to the siren.

To compare the same car made in different molds, refer to Plates 8 and 9. While at first glance the units appear identical, even to the color, the detailing is significantly different. Plate 8 is a metal version made in Japan, while Plate 9 is a Hong Kong unit in plastic. The quickest way to visually tell them apart is to look at the running boards. The Japanese units all have a metal plate on the top of these boards, while the Hong Kong versions omit this detail. If you can lift the units, another factor is quite obvious: the Japanese units are much heavier and have a very solid feel, while the plastic units almost seem to "float" in your hand and are comparatively flimsy. This "solid" feel is typical of all the Japanese units and is exemplified by the Stutz shown in Plate 4.

Hong Kong isn't alone in using one mold for different jobs. The Japanese stage coaches in Plates 37 and 38 are also made in the same mold, although they are trimmed out a little differently. The river boats shown in Plates 41 and 42 also differ only in name: the rest of the unit is identical (the "Mark Twain" unit pictured is missing one short smoke stack).

Another interesting point for the novice collector is the very fragile hood ornaments that are part of all the 1930s style autos shown in Plates 6-9 and in Plate 12. All have the same shortcoming, regardless of whether the cars are made in Hong Kong or Japan — namely, most of these cars that appear in the secondary market will have them broken off (and I've passed them up by the dozens just for this reason). In an attempt to solve this problem (and to make another different car), the 1931 "Classic" shown in Plate 10 features a grill without this fragile ornament. A careful look at this unit shows that it uses the same body mold as the 1931 Rolls, but does make use of this different style grill (although I know of no car in this period that uses a grill even remotely like this one).

Another item of interest is the Gran Prix formula one race car shown in Plate 22. It is perhaps more famous in Europe than in the U.S. due to the fact that the British John Player cigarette company was the first non-auto-related company to sponsor a Gran Prix race car. Their first car was very modest, only displaying a small sign advertising their "Gold Leaf" cigarettes. When the company introduced their new cigarette, known as the "John Player Special," this modest approach was abandoned, and their new car became a moving billboard using the new "logo" and color scheme to completely cover the car.

These same black and gold colors, and the distinctive logo shown on the "wings" of this car, also appear throughout London on the front of the "John Player Special" tobacco shops, a color combination that is very difficult to ignore — and often seems to clash with the neighboring shops. The car shown in Plate 23 is probably made from the same mold, but distributed by Radio Shack and uses their markings.

The high quality, typical of the Japanese units mentioned earlier, is not limited to the cars, but also appears in other items such as the "Mississippi" fire pumper shown in Plate 36. This unit not only has a very attractive appearance, with its bright red paint and brass-plated trim, but being mostly metal it also projects the same "solid" feel when you lift it.

Plate 53 shows a globe with a bakelite

base. This is quite unusual as most of these units use modern plastics rather than bakelite, which is normally associated with the 1930s.

China is also represented in this chapter. The 1930s style gas pumps shown in Plates 62-64 are very desirable. Several variations exist, but the one shown in Plate 62 is the only one in the series I have seen that appears to be a true commemorative unit that refers to the pumps' historical past.

NOTES

PLATE 1
ROLLS ROYCE (Circa 1912). Plastic Rolls with convertible roof. License plate number SD8451. Compare the body with the one shown in Plate 2. (See Text.) Hong Kong unit measures 9L x 5H.

PLATE 2
1912 SIMPLEX. This unit was in its original box, and marked as shown. The license plate number is 3288. This is a variation of Plate 1. (See Text.) Made in Hong Kong, and is 9L x 4-1/4H.

PLATE 3
MODEL "T" (Circa 1912). Japanese metal body and plastic trim unit distributed by Waco. Excellent detail with the controls concealed in the "Rumble Seat." It measures 9-3/4L x 6H.

PLATE 4
STUTZ BEARCAT (Circa 1914). Another
excellent unit from Japan. Metal body w/
plastic undercarriage. The controls are via the
spare tire at rear. The spare turns to tune,
while the "Knock-off" mounting bolt is the
volume control. This type of unit is
outstanding! 10-1/2L x 4-1/2H.

PLATE 5
STUTZ BEARCAT (Circa 1914). The hood
and gas tank are similar to the unit shown in
Plate 4. This one has a different style body,
and adds a convertible roof. Made in Hong
Kong and is 10-1/2L x 3-1/4W.

Courtesy BNZ

PLATE 6
L I N C O L N 1 9 2 8 M o d e l "L"
CONVERTIBLE. This is a Hong Kong unit
in plastic, but a Japanese version in metal is
also available. These have a very fragile hood
ornament that is often missing on used units.
(As does the 1931 Rolls.) I have passed up
dozens of these due to this fault. 10L x 3-3/4H.

PLATE 7
ROLLS ROYCE. 1931 Phantom II. Metal body w/BRASS plating. Has a plastic undercarriage. Outstanding unit, with excellent detail. Made in Japan, and is 10L x 3-3/8W.

PLATE 8
ROLLS ROYCE. As above, but in PEWTER. Note the metal trim on the running boards on the Japanese units. The controls on all the ROLLS are via the spare tires.

PLATE 9
ROLLS ROYCE. This is a Hong Kong unit made in plastic. While it is an excellent likeness, it lacks the detail of the Japanese units.

PLATE 10
1931 CLASSIC CAR. Hong Kong unit in plastic using the same body and design as the Rolls - but with a different grill that solved the broken ornament problem. (See Chapter introduction) 10L x 3-3/8W.

PLATE 11
GRILL DESIGNS. Front shot of the 1931 Rolls and 1931 Classic cars shown above. Both are in plastic, but the grill on the black car has a "Diamond" pattern, while the Rolls has the traditional grill and the very fragile hood ornament.

PLATE 12
DUESENBERG 1934 MODEL "J." Japanese unit w/metal body and plastic undercarriage. This unit has been painted due to its very poor condition. (Including a broken hood ornament). Possible BRASS versions exist. (This one was pewter.) Excellent detail. 10-1/4L x 3-1/4H.

PLATE 13
CADILLAC. Convertible (Circa 1963). Nice rendering in plastic from Hong Kong. The license is "CAD-1," and it is model number 7546. No distributer information appears on the box, or on the car. This one is grey, but a gold one is also available. The controls are on the bottom, near the front. It is 10L x 2-1/2H.

PLATE 14
THUNDERBIRD (Circa 1965). Marked "Philco Corp. Division Ford Motor Co. Model NT-11." Made in Hong Kong, and I'm told it was a "Give Away" for either buying a T-bird, or taking a test drive in one. (?) It comes in many colors, and measures 8L x 3H.

PLATE 15
MUSTANG FASTBACK. (Circa 1966). This Hong Kong unit is marked "Philco/Ford model P22". Probably used the same way as the T-Bird in Plate 14. Comes in other colors and measures 7-1/2L x 2-3/4W.

Courtesy WMC

PLATE 16
LINCOLN CONTINENTAL 1966 Model.
Distributed by Philco. This car possibly could
have been used the same as the T-Bird unit
shown in Plate 14. Made in Hong Kong, and
measures 8-1/2L x 2-1/2H.

Courtesy LCB

PLATE 17
MERCEDES BENZ SEDAN. Hong Kong
plastic unit, distributed by Trico. It measures
8-1/2L x 2-1/2H.

Courtesy BNZ

PLATE 18
BIG FOOT 4X4. AM/FM unit made in Korea
and distributed by ERTL (Vanity Fair). This
is based on an actual unit that follows the
competition circuit. It is 9-3/8L x 6-1/2H.

PLATE 19
FIRE CHIEF. Volkswagen Bug with Red light and Siren. The red light tunes the radio, while the volume is a thumbwheel on the bottom of unit. Made in Hong Kong, and is 6L x 4H.

PLATE 20
POLICE CAR. (Same details as Plate 19, but marked and colored differently.)

Courtesy BNZ

PLATE 21
RTL 208. Radio Tele Luxembourg sound truck. All die cast metal made in Great Britain and distributed by Corgi Toys ©1982. The controls are concealed behind the rear doors. 4-3/4L x 2-3/8H.

PLATE 22
JOHN PLAYER SPECIAL #2. Gran Prix racer with excellent detail. The controls are concealed under the drivers seat and behind the front wheel. (See chapter introduction.) Made in Hong Kong, and is 9-3/4L x 5W x 3H.

Courtesy BNZ

PLATE 23
SPEEDWAY SPECIAL #3. Hong Kong unit distributed through Radio Shack. This unit is identical to Plate 22, and is possibly based on that unit. Same details as above.

PLATE 24
PORSCHE RACER #7. British registration number 985866. Excellent detail with controls concealed in trunk. Made in Hong Kong, and is 8-1/2L by 2-3/4H.

PLATE 25
RADIO GOBOT. ©1985 by Tonka Toy.
Distributed by Playtime Products. This unit
converts into the robot shown in Plate 225.
Made in Hong Kong, and is 8L x 2-3/8H.

PLATE 26
ROBOCHANGE. Transformer unit in
automobile mode. (See Plate 254 for Robot
Mode.) ©1985 Tai Fong and made in Taiwan.
Comes with 3 different sets of decals to add
some variety to the unit. 7-3/4L x 2-1/4H.

PLATE 27
KNIGHT 2000. ©1982 Universal Studios.
Distributed by Royal Condor ©1984. Based on
the television series "Knight Rider." Korean
unit measures 9-1/2L x 2-1/2H.

PLATE 28
"E" TYPE JAGUAR. Made in Japan and distributed by Swank. This unit is missing the front bumper and windshield. The controls are the "bars" in the front wheels. 8-1/2L x 3H.

PLATE 29
"JEM" Glitter Gold Roadster. Made in Hong Kong, and distributed by Hasbro. Inc. ©1986. (The "Barbie doll" driver is not original.) This is a LARGE unit, measuring 24L x 9W x 7-1/2H.

Courtesy BNZ

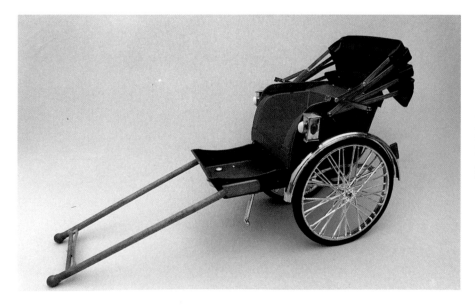

PLATE 30
RICKSHA. British design number 986010. Made in Hong Kong, and has controls via thumbwheels at your heels, if you were seated in this unit. (Barely visible in photo.) Unusual item, with good detail. Measures 13-1/2L x 7H.

PLATE 31
1864 LOCOMOTIVE. Pewter plated metal unit from Japan. The controls are on the smoke stacks. It is 9-1/2L x 4-1/4H

PLATE 32
C. P. HUNTINGTON LOCOMOTIVE. This model is based on the original unit housed in the California Railroad Museum. The controls are via knobs at top of "smoke stacks." Made in Hong Kong, and measures 11L x 5H.

PLATE 33
1864 IRON HORSE. Locomotive w/all metal construction. This unit is powered by 2 "C" cells inserted into the tubular main body. The controls are thumbwheels under the coal car. Made in Japan, and measures 10-1/2L x 4H.

PLATE 34
1826 LOCOMOTIVE. Plastic version, made in Hong Kong. The controls are on top of the smoke stacks, and it measures 9-1/4L x 4H.

PLATE 35
THE GENERAL. Locomotive made in Japan, but distributed through General Electric. Plastic unit measuring 9-1/4L x 7H.
Courtesy LCB

PLATE 36
1869 MISSISSIPPI. Fire Pumper w/all metal construction. Stamped metal base w/brass plated trim. Another Japanese unit with visible controls. (On top rear, behind large tank.) This unit is missing the two white hoses that rest in the metal brackets at the sides. 10-1/4L x 6-1/2H

PLATE 37
STAGE COACH. Overland Stage Express Co. Brass Plated wheels w/ red plastic body. Made in Japan, and has visible controls on the top. Note the U. S. Mail emblem embossed on the side door. This unit is 7-1/4L x 5H.

PLATE 38
STAGE COACH. Brown plastic version of Plate 37, using slightly different trim. The "Mail" logo is a metal plate, rather than being engraved into the plastic. Made in Japan, and is same size as above.

PLATE 39
CABIN CRUISER. Plastic w/metal trim unit from Japan. Outstanding details in lifeboat, radar and lamps. Possibly sat on base at one time. Made in Japan, and measures 11-1/2L x 6-1/2H.

PLATE 40
CABIN CRUISER. Another version from Japan. Mostly plastic w/metal trim. Nicely detailed and measures 8-1/2L x 3-1/2H (Boat Only).

Courtesy LCB

PLATE 41
MARK TWAIN. Riverboat sold at Disneyland. (And many other locations.) Plastic hull w/metal trim. Japanese unit with visible knobs on top deck. 12L x 7-1/4H

PLATE 42
BELLE OF LOUISVILLE. Riverboat made in same mold as above. I have also found a "TUPPERWARE BELLE" made from this same mold, so possibly other versions exist.

PLATE 43
1917 TOURING CAR. This **BRASS** plated metal unit is from Japan. The controls are via thumbwheels visible above the running boards. It measures 6-1/2L x 4-1/2H.

PLATE 44
1917 TOURING CAR. This **PEWTER** plated metal unit is identical to the one shown in Plate 43.

PLATE 45
1917 TOURING CAR. This white Japanese unit is made in plastic. It is the same design as the metal units show in Plates 43 & 44.

Courtesy WMC

PLATE 46
1908 TOURING CAR. This is a stamped metal car that sets on a plastic base, containing the radio. The base also has a presentation plate.

Courtesy BNZ

25

PLATE 47
GLOBE. Early unit from Japan, marked "Vista." Tuning is via the slider at side, while the volume control is at the top. It is 8H and 7 inches in diameter.

PLATE 48
GLOBE. Another unit from Japan, having a wide ring around the middle and marked "Star Lite High Sensitivity." It is 9-1/2H x 7 in diameter.

Courtesy BNZ

PLATE 49
GLOBE. Similar unit to Plate 47, but AM/FM version with tuning via knob at top; volume via thumbwhell at side. Made in Japan, and same size as above.

Courtesy BNZ

PLATE 50
GLOBE. Japanese unit w/metal rings and seahorse mountings on a plastic base. The globe itself is wood, with a heavy paper covering. Distributed by Heritage and is 5-1/2H with 4-1/2 diameter globe.

PLATE 51
WORLD GLOBE AM RADIO. Distributed by Radio Shack. Modern world globe on an unusual shaped base. The box features a young girl studying her geography while listening to the radio. Made in Korea and stands 8 inches high with a 5 inch globe.

PLATE 52
GLOBE. All wood construction, and probably an early unit from Japan. The controls are visible, and it is 8-1/2H x 5-5/8W.

PLATE 53
GLOBE. Old World type w/signs of the Zodiac around the rim. This wooden globe sets on a bakelite base which contains the radio. Made in Hong Kong, and is 10H and the globe is 7-3/4 inches diameter.

PLATE 54
GLOBE. Modern world version made in Japan and distributed by Omsco Lite. This is an early 2 Transistor unit. The Globe is 5-1/4 in diameter, and the base is 4-1/2 inches in diameter by 1-1/2H.

PLATE 55
SANTA MARIA. Plastic body w/brass trim. This unit was made in Japan. The base has a presentation plate to be engraved for an award or other occassion. It measures 11-1/2L x 11-1/2H.

Courtesy LCB

PLATE 56
SHIPS TELEGRAPH. Metal construction w/painted base and BRASS plating. Has plastic face plate with speeds and frequency around edge. Tuning is via speed control, volume via thumbwheel in base. Made in Japan, and is 10-3/4H x 4-1/4W.

PLATE 57
SEA WITCH. 1864 Clipper Ship, Made in Hong Kong, and distributed by Windsor. This unit sets on a base that contains the radio. It is 13L x 13H.

PLATE 58
FRIGATA ESPANOLA ANO 1780. Another version of a sailing ship using the same concept as Plate 57. Made in Hong Kong and measures 12L x 12H.

Courtesy BNZ

PLATE 59
JAGUAR GRILL. This is a little different item for the automobile buff, and illustrates how piece parts of autos can be made into radios. This one features the grill and "Logo" of the Jaguar. There is another version using the grill and distinctive "Star" of the Mercedes-Benz.

Courtesy LCB

PLATE 60
LIFE PRESERVER. Clock/Radio Model BCR1300. Distributed by Aimor Corp. L. A. Calif. Made in Japan, and is 10-1/2 in diameter.

PLATE 61
TWA TRANSWORLD 747 JET. This is an excellent likeness of
the famous Jumbo Jet. Made in Hong Kong and distributed
through Windsor. It is 13L with a 12-1/2 inch wingspan.

Courtesy LCB

PLATE 62
GAS PUMP. 1930s Style unit made in China and distributed
by Synanon. It is an **AM/FM** unit and a decal at the rear states:
"This historical Gasoline pump was used during the 1930s by
Standard Oil Company of California, the predecessor of
Chevron Corporation." It is 9-1/2H x 3-1/8 Diameter.

PLATE 63
GAS PUMP. Same as Plate 64, but a "Generic" type using
only the trade mark of the "Ethyl" Corp.

30

PLATE 64
1930 GAS PUMPS. These are all AM/FM units
made in China. The tuning is via the "Gauge"
on the front; volume via a knob at rear. The
Standard Oil one is marked "Historical Version
Used during the 1930s." (See Plate 62). All are
3-1/8 Diameter by 9-1/2H.

PLATE 65
MATT TRAKKER. Rhino Rig. Box marked "©Mask trade mark
of Kenner Parker Toys Inc. 1985." Distributed by Playtime
Products Inc. Number 011505. Made in Hong Kong, and is
7-1/4L x 4H.

NOTES

Media Stars & Heroes

This chapter consists of characters that owe their fame to the media. Their roots trace to comic strips, children's books, radio programs, movies, or television. These heroes and stars are fictional, as opposed to the ones pictured in the next chapter: "Character and Figural."

It is through this section that many of the collectors of tomorrow will be born. Toys and novelties based on cartoon or comic heroes are a link to our youth. It may seem strange to today's collector that a radio like the "Pound Puppy" or "He Man" could ever have appeal, but they are the heroes of our children (or grandchildren). Yes, they will be collectible for the same reason most things are: the nostalgia factor, and the simple fact that the common radios of today will not be plentiful in the future.

The famous media heroes of previous years — Jack Armstrong, Captain Mid-night, The Green Hornet and The Lone Ranger — will probably never appear as transistor radios (although the Lone Ranger has been made into a vacuum-tube unit). The only heroes of my long-lost youth that have this honor are the ones made famous by Walt Disney, mainly Mickey Mouse and his many friends. These Disney characters have been favorites since radio began, and Mickey has appeared for every generation as a radio shortly after his birth in 1928. (The 1933 Emerson is the most collectible. It's now worth about $1000.) Other Disney characters, such as Snow White and the Three Little Pigs, have been made into vacuum-tube radios, but I have yet to find them as transistorized units.

A large portion of this chapter continues the Disney legend. Many of his characters are represented here, including some lesser stars like the Wuzzles, Winnie-the-

Pooh, and Goofy. True Disneyana buffs look for items marked "Copyright Walt Disney" or "Copyright Walt Disney Enterprises." These marks precede our radios by many years, but one word of caution: the other item normally associated with Mickey Mouse in the 1930s is the famous "Pie-Eyes," and these *do* appear on these late transistor radios! A good example is the reclining Mickey shown in Plate 113. This Mickey figure happens to be detachable by removing two clips inside of the radio. I have seen this "doll" in several different dealer locations, each one having a very high price on it (one had it for $150). When I tried to explain the source of this "doll," the dealer was very skeptical, assuring me that NO Mickey item has had "Pie-Eyes" since 1938 (Disney buffs, take care!).

Many of the character radios manufactured today are generated by television exposure. While Snoopy, Spider Man, Bugs Bunny and Batman trace their roots to other media, they were made worthy of a radio due to television. A good example of this is the Sesame Street units shown in Plates 139 through 149. The radios are based on Jim Henson's Muppets, but owe a portion of their fame to the PBS series "Sesame Street." The logo for this series has the letters "CTW" as part of the street sign, and means "Children's Television Workshop," referring directly to the television roots.

The radios pictured feature most of the characters that appear on the series, with the exception of Grover. In the children's books of this series, Grover appears more frequently than the Cookie Monster, who IS represented — although I have yet to find a radio featuring Grover, even in the "group" shots. The omission of a few characters appears to be typical of units that appear as a series. The "Wuzzles," ©Walt Disney, have many different animals pictured in the books, yet I have found only two as radios. The "Care Bears" and "Cousins" have many different characters, but again, I have found only

three as radios. Perhaps it isn't logical, and probably not profitable, to make each and every character in a series into a radio.

Other media radios, like "He-Man" and the "Dukes of Hazard," also trace directly to television programs without any other media involved. (Indeed, the character is often on TV, then appears as a comic strip.) A look at Saturday morning TV will yield a group of heroes that may not be familiar to the older collector. These new heroes with strange names, such as Thunder Cats, Ninja Turtles, and Mighty Thor, have replaced the Jack Armstrong and Captain Midnight of yesteryear.

Thankfully, some of my old favorites — whose roots are in movie cartoons — are also represented. Tom and Jerry, Popeye, and Bugs Bunny are among them. (So are Mickey and Donald — but they have appeared in EVERY media device.)

Some older stars, like Annie, are also represented, but this radio is based on the movie version of the little "Red Head," not her years as a radio and comic-strip star. Oddly enough, some lesser heroes (those who never were television stars) are also present: Little Lulu, Raggedy Ann and Andy, and even Holly Hobbie appear, proving you *can* be famous without television.

Sadly, since most of these radios were made for children, some of them lack the detail and design effort of other units in this book. Some are almost two dimensional, using a flat outline of the character with a paper decal or a simple spray-painted front. Others, like the Sesame Street group, are three-dimensional and reflect better workmanship.

Several radios in this section are "Sing-A-Long" types (Plates 75, 120, etc.). This feature allows the user to tune in to their favorite "pop" station, wait for the disc jockey to play the current hit, then switch in the microphone and sing along with the recording star. Untold future generations of Michael Jacksons and Madonnas have been inspired by this novel feature. Most of the radios purchased on the secondary market will be missing these microphones (and I suspect many were "lost" by the parents. How many times can you listen to a child scream, "I'm bad, bad, bad?"). This is a good place to bring up Rule 1 about the secondary marketplace: "Anything not firmly attached to the unit will be missing!" It is a rare child (or adult, for that matter) who can keep these loose items with the radio over a period of years.

This same rule made this chapter hard to caption. Most of these radios have two copyrights: one for the character, and one for the company using this character to make the radio. A large number of manufacturers place this copyright information only on the battery compartment cover — an item with an unbelievable loss rate in the secondary market! In spite of this handicap, every effort was made to identify copyright information, even referring to other sources when available.

NOTES

PLATE 66
THE "A" TEAM (B. A. Baracus). Based on the televison series of the same name. "©1983 Cannell Products," are the only markings, but probably should have a studio copyright also. Made in Hong Kong, and is 5W x 4-3/4W.

Courtesy LCB

PLATE 67
ANNIE AND SANDY. This unit is based on the movie version of "Annie," not her years as a comic strip hero. ©1981 Tribune Company Syndicate Inc. — Columbia Pictures Industries Inc. Distributed by Prime Designs. Made in Hong Kong, and is 4-1/2H x 4-1/2W.

PLATE 68
BATMAN. Probably based on the television series of the 1960s. ©1978 National Periodic Publications. Made in Hong Kong, and is 5-1/2H x 6W.

Courtesy BNZ

PLATE 69
BOZO THE CLOWN. Another favorite that traces his fame to television. This unit is made in Hong Kong, and ©Larry Harmon Pictures Corp. It is distributed by Sutton Associates as an authorized user. It measures 6-3/4W x 5-3/4H.

Courtesy BNZ

PLATE 70
BARBIE RADIO SYSTEM. AM/FM unit with twin speakers for stereo-like sound. The main center unit features a 2D outline with a paper decal of BARBIE. The main unit is 6 x 6, and each speaker is 5-1/4 x 5-1/4. ©1984 Mattel Inc. and distributed through PowerTronic by Nasta (Model 20041). Made in Hong Kong.

PLATE 71
BUGS BUNNY. Eating his famous carrot, and you can almost hear him say "Ehh, Whats Up Doc?" Nice unit from Hong Kong. ©Warner Brothers (No Date). It measures 5-1/4H x 4-1/4W.

Courtesy WMC

PLATE 72
BUGS BUNNY. Another version of his classic "Whats Up?" pose, but this one is three dimensional. Made in Hong Kong, and is 6-1/2H x 3-3/4 in diameter.

Courtesy WMC

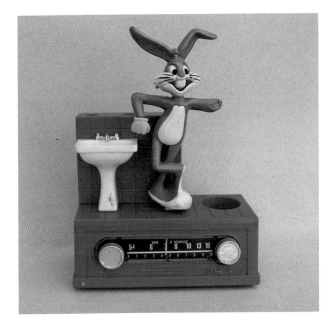

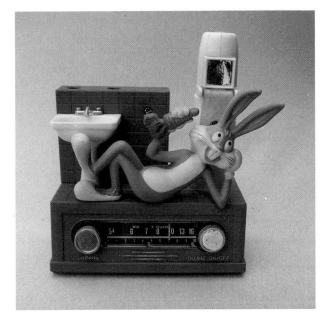

PLATE 73
BUGS BUNNY. This is a toothbrush holder, missing the
toothbrush. It would rest against "Bugs" stubby arm. A similar
unit featuring "Raggedy Ann" is shown in Plate 134. Nice unit
to teach children to brush their teeth. Made in Hong Kong, and
is 8-1/2 x 6W.

Courtesy BSG

PLATE 74
BUGS BUNNY. A different version of a toothbrush holder
showing a reclining Bugs with his carrot. Made in Hong Kong and
is 6W x 6H.

Courtesy of BNZ

PLATE 75
BUGS BUNNY. Pointing to station. This "pointing" technique
is used in several radios in this section. These radios are also "sing-
a-long" types. A Hong Kong unit measuring 7-1/4W x 8H.

Courtesy WMC

PLATE 76
CARE BEAR COUSINS. The "Cousins" are actually lions. 2D
unit w/slider controls. ©1985 American Greeting Cards and radio
distributed by Playtime Products ©1983. Made in Hong Kong and
is 5H x 3W.

PLATE 77
CARE BEARS/COUSINS. Family reunion with a cousin on the left, Funshine in center, and Cheer Bear on the right. All ©1985 American Greeting Cards and distributed by Playtime Products Inc. Made in Hong Kong and measure 5-1/2H x 5-1/2W.

PLATE 78
CARE BEARS. A 2D unit showing Cheer Bear on the front and includes the words, "Care Bears" on the rainbow. ©1983 by American Greeting Cards and distributed by Playtime Products Inc. ©1987. Made in Hong Kong and is 4-3/4 in diameter.

PLATE 79
CARE BEARS. A 2D unit with a very glossy decal on the front, but omits the words "Care Bears." Details same as Plate 78.

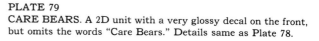
Courtesy BNZ

PLATE 80
CABBAGE PATCH GIRL. Based on the cartoon characters and ©1985 by Playtime Products. Hong Kong units that measure 6-1/4L x 5-1/2H.

Courtesy WMC

PLATE 81
CABBAGE PATCH BOY. Same as Plate 80, but measures 6-1/4L x 5-1/4H.

Courtesy WMC

PLATE 82
CABBAGE PATCH KIDS. A two dimension unit featuring slider controls. ©1983 Original Appalachian Artworks and distributed by Playtime Products ©1981. Made in Hong Kong, and is 5H x 3W.

PLATE 83
DONALD DUCK. Two dimensional unit, ©Walt Disney Productions and distributed by Philgee International. Made in Hong Kong, and is 7-1/2H x 5W.

PLATE 84
DONALD DUCK. Similar unit to Plate 83, but has loud speaker behind Donalds mouth. ©WDP and distributed by Philgee International. Made in Hong Kong and is 7H x 4-1/4W.

Courtesy BNZ

PLATE 85
THE DUKES OF HAZARD. Based on the televison series of the same name. Two dimensional unit with paper decal showing the stars and their car. Marked "T. M. indicated Trademark of Warner Bros. Inc. ©1981." It is distributed by Justin Products and made in Hong Kong. It measures 7-1/2L x 4-1/4H.

PLATE 86
FRED FLINTSTONE. You can almost hear him say "Yaba Daba Doo!" Based on the television character and ©1972 Hanna Barbera Productions. Sutton Associates Ltd. as an authorized user, and is also marked "TM Columbia Pictures." Made in Hong Kong, and is 6-1/2H x 7W.

PLATE 87
THE FONZ. The words "Happy Days" appear on the side of this unit, leaving little doubt as to its origin. ©1977 Paramount Pictures Corp. Made in Hong Kong and is 6H x 4-1/2W.

PLATE 88
GARFIELD WITH ODIE CHARM. Americas favorite cat in a
nice rendering. The Odie charm is frequently missing on used
units. ©1978 United Features Syndicate Inc. and distributed by
Durhan Industries Inc. Made in Hong Kong, and is 4W x 3-3/4H.

Courtesy LCB

PLATE 89
GARFIELD "MUSIC IS MY LIFE" By Jim Davis. Different
rectangular unit with molded Garfield. ©1978 United Features
Syndicate Inc. and distributed by Durham Industries Inc. (Model
3600). Made in Hong Kong, and is 5H x 3W.

PLATE 90
GOOFY ON WAGON. This unit is also a bank. I doubt if it is an
authorized version, as the artwork doesn't appear to be that of
Disney. Made in Hong Kong, and is 7H x 7L.

PLATE 91
GUMBY AND POKEY. Rectangular AM/FM unit showing
Gumby riding Pokey. Distributed by Lewco (Model 7012). Made
in China and is 6H x 3W.

43

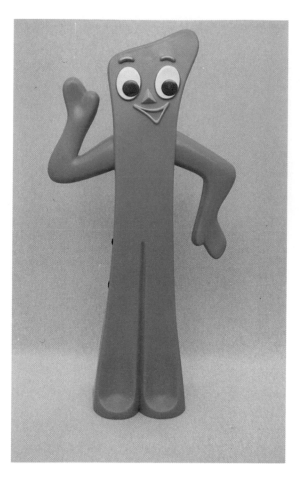

PLATE 92
GUMBY. This 12 inch Gumby is ©1985 by Perma Toy and distributed by Lewco (Model 7015). It is an AM/FM unit made in Hong Kong and measures 12H x 6W.

PLATE 93
HOLLY HOBBIE. "Start Today on a Happy Note." Nicely shaped 2D unit. Holly items are very collectable, and somewhat hard to find. Marked "©American Greetings Corp. - Knicker Bocker Toy Co. Inc." Made in Taiwan it is 8H x 4-3/4W.

Courtesy LCB

PLATE 94
HOLLY HOBBIE. This little charmer is leaning on her 1930s cathedral style radio. ©American Greetings Corp. Made in Hong Kong, and is 7-1/2H x 6W.

PLATE 95
HOLLY HOBBIE. Somewhat rare unit of Holly in a rocking chair. Also unusual in that the base is round. ©American Greetings Corp. and distributed through Vanity Fair. Made in Hong Kong, and is 7-1/2H x 4 inch diameter.

Courtesy LCB

PLATE 96
HE-MAN/SKELETOR. Two sided radio featuring He-Man on one side, and Skeletor on the other. ©1984 Mattel and distributed by Nasta (PowerTronic). Made in Hong Kong, and is 5H x 4W.

PLATE 97
SKELETOR. This is the other side of the radio pictured in Plate 96.

PLATE 98
HE-MAN. Unusual 3D version of the Saturday cartoon hero. ©1984 by Mattel and distributed through Nasta Inc. Made in Hong Kong, and is 5H x 3-1/2W.

Courtesy LCB

PLATE 99
HUCKLEBERRY HOUND. A cartoon favorite for all ages. This is ©Hanna-Barbera Productions and is distributed by Markson's Radio of Chicago, Ill. Made in Hong Kong, and is 6H x 5W.

Courtesy LCB

PLATE 100
HUCKLEBERRY HOUND AND YOGI BEAR. Two of the all time favorites of the Saturday morning crowd. ©Hanna-Barbera Productions Inc. and distributed through Markson's Radio, Chicago Ill. Made in Hong Kong, and is 5-1/4W x 4-1/2H.

PLATE 101
KERMIT THE FROG. This unit is about the cutest of the 2D types. Really nice detail and coloring (I like Kermit!). ©1984 Henson Associates and distributed by Nasta. Hong Kong unit measures 5W x 4-1/4H.

Courtesy LCB

PLATE 102
LITTLE LULU. A favorite of the older generation, but somewhat limited appeal for today's children. Rare unit ©Western Publishing Co. Inc. and licensed to J. Swedlin Inc. Made in Hong Kong, and is 7H x 5-1/4W.

Courtesy BNZ

PLATE 103
"SCOOBY DOO". Another cartoon favorite ©Hanna-Barbera Productions Inc. 1972. Sutton Associates Ltd. as an authorized user. Made in Hong Kong, and is 7-1/2H x 4W.

Courtesy WMC

PLATE 104
MASTERS OF THE UNIVERSE. A Saturday morning cartoon favorite. Distributed by Nasta (Model No. 21001). Hong Kong unit measuring 5H x 4-1/2W.

PLATE 105
MARSHMELLOW MAN. Based on the movie "Ghost Busters." The base of this radio is similar to the unit shown in Plate 148, and uses the lamp as a night light. ©1984 Columbia Pictures and distributed by Concept 2000 (Model GB1085). Made in China and is 7-1/2H x 6-1/2W.

PLATE 106
MICKEY MOUSE IN CAR. This radio has universal appeal. The upward looking Mickey seems to look right into your eyes as you hold the unit. ©WDP and distributed by Concept 2000 (Model 181). Made in Hong Kong, and is 6-1/4L x 4-1/4H.

Courtesy LCB

PLATE 107
MICKEY MOUSE. Nice rendering of the "pie eyed" Mickey in a night-lite radio. The control on the left contols the brightness of a lamp behind Mickeys face. ©WDP and distributed by Concept 2000 (Model 402). Made in Hong Kong, and is 8W x 6H.

PLATE 108
MICKEY MOUSE. Mickey is hitting some "hot licks" with this sing-a-long unit in the shape of a guitar. Made in China, and measures 11-1/2L x 5-3/4H. Concept 2000 (Model WD-1003) ©W.D. Co. (No Date)

Courtesy BNZ

PLATE 109
MICKEY MOUSE. This two dimensional unit is ©WDP and is distributed by Concept 2000 as well as Philgee International. Made in Hong Kong, and is 6-3/4H x 6-5/8W.

PLATE 110
MICKEY MOUSE. 2D unit, posed similar to Plate 109 — But note the base treatment. He is actually resting on his chin and nose! ©WDP and distributed by Philgee International. Made in Hong Kong, and is 6W x 5-1/2H.

Courtesy BNZ

PLATE 111
MICKEY MOUSE. Another "big eared" version that is somewhat unusual in its treatment of the base. Mickey rests on two molded bars instead of a full base. ©WDP and distributed by Philgee International. Made in Hong Kong, and is 5W x 3-3/4H.

Courtesy LCB

PLATE 112
POINTING MICKEY. Another of the "pointing" series, but note the "Pie Eyes" in this rendering of Mickey. These normally appear only in the 1930s versions of the world's most famous mouse. Made in Hong Kong, and is 7-1/2W x 7-3/4H.

PLATE 113
RECLINING MICKEY. This "Pie Eyed" Mickey is three dimensional, and removable! (See chapter introduction for more details about this "Doll.") This radio is often missing the doors. Made in Hong Kong, and distributed by Concept 2000. (No Model Number). It is 8-1/2H x 8W.

PLATE 114
MICKEY MOUSE. This is a pendant radio with a 29 inch chain, and designed to be worn around your neck. ©WDP and distributed by Concept 2000 (Model 262M). It is a Hong Kong unit that is 3-1/2 in diameter.

PLATE 115
MICKEY MOUSE. Early 2 transistor unit from Japan. The controls and markings are on the rather large ears. ©Walt Disney Productions and distributed by Gabriel. 6-1/2W x 6-1/2H.

Courtesy BNZ

PLATE 116
MICKEY MOUSE. This was the first of the square "Big Ear Tuning" units. ©WDP and distributed through Radio Shack in 1987. Another unit showing Mickey in 3/4 view was sold in 1988/89. Made in Hong Kong and is 3-1/4 square.

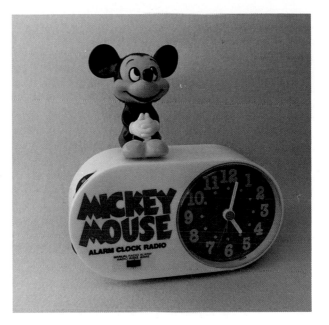

PLATE 117
MICKEY MOUSE. In arm chair. This unit is made of a soft plastic and shows Mickey "laid-back," and probably watching his re-runs on television. ©WDC and distributed through Concept 2000 (Model WD1043). Chinese unit measuring 4-3/4H x 4-1/4D.

PLATE 118
MICKEY MOUSE ALARM CLOCK RADIO. Features a very attractive Mickey on the top. ©WDP and distributed through Concept 2000 (Model 409). Made in Hong Kong, and is 8L x 8-1/2H.

PLATE 119
MICKEY AND DONALD. Nicely detailed unit in a childs nite lite. The dimmer controls the crystal ball. ©WDP and distributed by Concept 2000. Made in Hong Kong, and is 8H x 7W.

PLATE 120
MICKEY MOUSE SING-A-LONG. Mickey, Donald, and Pluto riding the band wagon. The microphone, that plugs into a cord at the rear, is shown between Mickey and Pluto. ©WDP and distributed by Concept 2000. Made in Hong Kong and is 7-1/2H x 8L.

PLATE 121
MICKEY AND MINNIE AT MUSIC CITY. Nice rendering of the famous couple in a Chinese made unit. No marks appear on the radio giving Walt Disney credit. It is 8-1/2H x 6-1/2W.

Courtesy LCB

PLATE 122
PINK PANTHER. This cartoon favorite is made in Hong Kong and marked "©Talbot Toys 1982 - Model 3005." It measures 6H x 3-1/2W.

Courtesy WMC

PLATE 123
MORK FROM ORK EGGSHIP RADIO. Based on the popular television series. ©1979 Paramount Pictures and distributed by Concept 2000 (Model 4461). Hong Kong unit that is 7H x 4-1/4 in diameter.

PLATE 124
MY LITTLE PONY. ©1973 Hasbro Industries and distributed by Durham Industries Inc. Made in Hong Kong, and is 4W x 3-1/2H.

Courtesy BNZ

PLATE 125
MY LITTLE PONY. Rectangular unit with molded pony on the front. Similar construction as the "Garfield" unit shown in Plate 89. ©1983 Hasbro Industries and distributed by Durham Industries Inc. Made in Hong Kong and measures 5H x 2-3/4W.

PLATE 126
PACMAN. This cute little fellow became famous as an arcade game that took the country by storm. (A slightly larger version of this same shape has been made into a telephone.) ©1982 Bally Midway Mfg. ©1982 Tiger Electronic Toys. Made in Hong Kong, and is 4-1/4 in diameter.

PLATE 127
POPEYE. An older character, but still popular due to the many cartoons in syndication. ©King Feature Syndicate and distributed by Philgee International. Made in Hong Kong, and is 6-1/2H x 7W.

Courtesy WMC

PLATE 128
POOCHIE RADIO SYSTEM. This AM/FM
unit features twin speakers for stereo-like
sound. It is similar to the "Barbie" unit shown
in Plate 70.

Courtesy LCB

PLATE 129
POOCHIE. Another favorite of the young
girls. Nice 2D unit. ©Mattel Inc. 1983. Made
in Hong Kong, and is 5-1/4W x 3H.

PLATE 130
POUND PUPPY. The radio is in the molded
plastic base. The puppy is made of soft cloth.
This was sold through Radio Shack stores
during the Christmas season of 1987. The base
unit is made in Hong Kong, while the puppy
is made in China. It measures 7-1/4W x 5-1/2W.

PLATE 131
RAGGEDY ANN AND ANDY. One of the better 2D units featuring an embossed paper decal on the front. This is an improvement over most of paper label types. ©1973 Bobbs Merrill Co. Inc. and distributed by Philgee International Ltd. Made in Hong Kong, and measures 6-1/4W x 6-1/2H.

PLATE 132
RAGGEDY ANN AND ANDY. Heart shaped 2D unit with paper decal on front. ©1974 Bobbs Merrill Co. and distributed by Philgee International. Hong Kong unit measures 4-3/4W x 4H.

PLATE 133
RAGGEDY ANN SING-A-LONG. Also featuring the "pointing hand" of some other units. Note that the microphone is in the shape of a tulip! ©1975 Bobbs Merrill Inc. Distributed by Concept 2000 (Model 458). Made in Hong Kong, and is 8-1/2W x 7H.

PLATE 134
RAGGEDY ANN. Toothbrush holder and radio. This is the same concept as the Bugs Bunny one shown in Plate 73. Made in Hong Kong, and is 8H x 6-1/4W.

Courtesy LCB

PLATE 135
PINOCCHIO. Although a 2D unit, it features nice detail and good coloring. Pinocchio is comparativly rare, and I know of no other unit that features him. Made in Hong Kong, and is 6-3/4H x 6W.

PLATE 136
POPPLES. An unusual radio, as it has hair (although the color is pink,) felt ears, and a two-tone tail made of fabric. Most radios of this type mold these items into the plastic, rather than make them external. (Probably will have a high fatality rate in used units.) ©American Greetings Corp. with Playtime Products as an authorized user. Made in Hong Kong, and is 5-1/2H x 5W.

PLATE 137
PRINCESS OF POWER CASTLE. Something for the gentle sex to identify with. Another Saturday cartoon heroine. ©1985 by Mattel and distributed through Nasta. Made in Hong Kong and is 7-1/2H x 4-1/4W.

Courtesy LCB

PLATE 138
RAINBOW BRITE. This AM/FM unit is ©Hallmark Cards and distributed through Vanity Fair by ERTL. A Korean unit that measures 6H x 4-1/4W.

PLATE 139
SESAME STREET. Bert in tub with his rubber duck. This duck is not attached and is usually missing in used units. The logo "CTW" that appears above the name means "Childrens Television Workshop." It is ©Muppetts Inc. and measures 6L x 6H.

PLATE 140
SESAME STREET. Bert, Ernie, Oscar, Big Bird, and the Cookie Monster under the street light. One of the few units that show most of the gang. ©Muppets (No Date). Made in Hong Kong, and is 9-1/2H x 5W.

PLATE 141
SESAME STREET. Bert has obviously just been awakened by the alarm clock and is ready to greet the new day with a smile on his face. Distributed by Concept 2000 (Model 4800), and made in Hong Kong. It is 7H x 8L and the clock is 7-1/2 inches in diameter.

Courtesy BNZ

PLATE 142
SESAME STREET. Big Bird leaning on a 1930s style radio. AM/FM unit, with the box marked ©1985 Muppets Inc. Distributed by JPI of New York. There are no markings on the radio except, "Made in China." It measures 6-1/2W x 7-1/2H.

PLATE 143
SESAME STREET. Big Bird sing-a-long Band. Bert, Ernie, and Big Bird making cool notes. This unit has a cord in the rear that plugs into the detachable black microphone, allowing you to sing along with your favorite song. (Since the mike is detachable, it is often missing.) ©1971, 1977 Muppets Inc. and distributed by Concept 2000 (Model 4701). Hong Kong unit measures 10H x 5-1/4W.

PLATE 144
SESAME STREET. Big Bird in 2D version with paper decal. ©1985 Muppets Inc. and distributed by Concept 2000 (Model 1720). Made in Hong Kong, and is 5H x 4W.

PLATE 145
SESAME STREET. Big Bird on his nest. ©Muppets Inc. Hong Kong unit that is 7-3/4H x 4-3/4 in diameter.

PLATE 146
SESAME STREET "COOKIE TIME
CLOCK RADIO." The Cookie Monster is
busy baking his favorite food, although he
appears to be ready to drown in them. Nice
unit, featuring one of lesser Sesame Street
characters. Made in Hong Kong, and is
9-1/2L x 5-3/4W.

Courtesy BSG

PLATE 147
SESAME STREET. Oscar The Grouch in his trash can. Very
accurate rendering of the classic drawings, even to the dent in the
side. ©Muppets Inc. (No distributor information). Made in Hong
Kong, and is 7H x 3-1/2 in diameter.

PLATE 148
SESAME STREET CENTER STAGE. Bert and Ernie do a soft
shoe on a light up stage. The light is controlled by the slide switch
in the very center. A small bulb, shielded by the yellow reflector
just above the switch, provides the foot lights. ©1985 Muppets Inc.
and distributed by Concept 2000 (Model 1723). Made in Hong
Kong, and is 6-1/2W x 6-1/2H.

PLATE 149
SESAME STREET. Bert and Ernie as pals. ©1985 Muppets Inc. and is 7-1/2H x 5-1/4W.

PLATE 150
SIX MILLION DOLLAR MAN. This "back pack" radio is actually a crystal set. (And technically doesn't belong in this book.) Since it has been marked down twice, finally to 49 cents, it probably wasn't all that popular. ©1973 Universal Studios and distributed through Kenner.

Courtesy LCB

PLATE 151
SNOOPY ON DOGHOUSE. This is THE classic pose of the worlds most famous pooch. Snoopy lifts up to become the handle for this radio. The controls are neatly tucked up under the eves of the house. ©1958 United Features Syndicate Inc. and distributed by Determined Products. Made in Hong Kong, and is 6-3/4H x 4W.

PLATE 152
SNOOPY OUTLINE. This is one of the more common Snoopy items you will find. Most have the paint worn off the edges, although it is easy enough to repaint. ©1958 - 1974 United Features Syndicate. (No distributor information). Made in Hong Kong and is 6-1/2H x 5W.

PLATE 153
SNOOPY POINTING. Another in the "pointing" series, although this one includes the sing-a-long mike. The unit has no trade mark information on it anywhere. The only marks are on the rear, and they say "Hong Kong." 7-1/2H x 7W.

PLATE 154
SNOOPY AND WOODSTOCK ON ROCKET. Off to the moon. Woodstock seems to be the pilot; while Snoopy, in his space suit, appears to be a mere passenger. The controls for the radio are via the exaust pipes. ©1958, 1965 United Features Syndicate Inc. Distributed through Determined Products Inc. Made in Hong Kong, and is 9-3/4L x 7-1/2W.

PLATE 155
SHIRT TALES. Nice 2D unit with lots of action and nice colors. This is ©1980, 1981, 1982 by Hallmark Cards. It is distributed by Royal Condor. Made in Hong Kong, and is 7-1/4L x 4H.

Courtesy LCB

PLATE 156
SMURFETTE AND SMURF. Pair of matching radios showing two different Smurf figures. Both units are made in Hong Kong and measure 5H x 2-3/4W. The Smurf is ©1981 and the Smurfette is ©1982 by Nasta Inc.

PLATE 157
CHARLIE BROWN, SNOOPY AND WOODSTOCK. A 2D unit with a paper decal. ©1958 United Features Syndicate and has no distributor information. Hong Kong unit measures 7H x 5-3/4W.

PLATE 158
THE SMURF. This unit is ©Peyo 1982. Licensed by Wallace Berrie Co. and distributed by Nasta Industries Inc. On top of all that, it is made in Hong Kong! 4H x4W.

PLATE 159
SPIDERMAN. Another "radio system" based on the "Secret Wars" series. This unit is similar to the "Barbie" and "Poochie" ones elsewhere in this section. AM/FM unit, made in Hong Kong, and is 7-3/4H x 5W.

Courtesy LCB

PLATE 160
SPIDERMAN. This unit is part of the "Secret Wars" series, and looks somewhat like a Frisbee. ©1984 Marvel Comics and distributed by PowerTronics/Nasta. Made in Hong Kong, and is 5 inches in diameter.

PLATE 161
SPIDERMAN HEAD. A nice 3D unit with thumbwheel controls at the base. It is ©1978 by Marvel Comics Group, and is distributed by Amico Inc. Made in Hong Kong, and is 5-1/4H x 3-1/4W.

PLATE 162
SUPERMAN. Exiting Phone Booth. Really neat unit, although when comparing it with the Superman shown in Plate 163, the face details could stand some improvement. ©1978 Vanity Fair and D. C. Comics. Made in Hong Kong, and is 7H x 3W.

Courtesy LCB

PLATE 163
SUPERMAN. The "Man of Steel" in a 2D rendering. Nice unit, with bright colors and good detail. ©National Periodical Publications 1973. Hong Kong unit that measures 5H x 6W.

Courtesy LCB

PLATE 164
TRANSFORMERS. This rectangular unit features a Transformer figure with movable arms and legs. ©1984 Hasbro Industries Inc. and distributed through Nasta Inc. Made in Hong Kong, and is 5H x 2-3/4W.

PLATE 165
WRINKLES. The friendly dog on a pedestal base. ©1986 Playtime Products Inc. Made in Hong Kong, and is 7-1/2H x 6W.

Courtesy LCB

PLATE 166
STRAWBERRY SHORTCAKE. This radio is similar to the Cabbage Patch Kids unit shown in Plate 82. It features Strawberry playing her Guitar. ©1983 American Greetings Inc. and distributed by Playtime Products ©1984. Made in Hong Kong, and is 5H x 3W.

PLATE 167
STRAWBERRY SHORTCAKE. "Music Makes The Whole World Glow." Strawberry is baking in her "light-up" oven. A small switch on the side controls a lamp inside the oven. ©1982 American Greeting Cards and distributed by Justin Products Inc. (Procision Model 456). Hong Kong unit measures 7-1/2L x 7-1/2H.

PLATE 168
"SWEET SECRETS" This unusual looking item is really a young girls delight. When you open the compartments, a complete make-up studio is revealed — with mirror, chairs and all the various items every girl needs. It is ©1985 by Lewis Galoob Toys and made in Hong Kong.

Courtesy BNZ

PLATE 169
"SWEET SECRETS". This is the unit when it is opened. Note the fine detail in the chairs and sink.

Courtesy BNZ

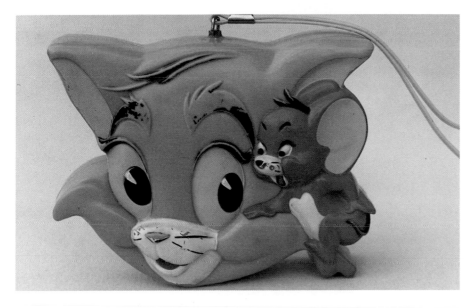

PLATE 170
TOM AND JERRY. One of the rarer units, and somewhat unusual as two characters are molded together in three dimensional presentation. These two are old favorites of the cartoon set. ©1972 MGM and distributed by Marx. Made in Hong Kong, and is 6W x 4-3/4H.

Courtesy LCB

PLATE 171
WUZZLE. This is "ButterBear," a cross between a Butterfly and a Bear. ©WDP and distributed through PowerTronics by Nasta. Made in Hong Kong, and is 7-1/4H x 5W.

PLATE 172
WUZZLE. This one is called "Bumble Lion, a cross between a Bumblebee and a Lion. Other Wuzzles exist, but these are the only two I have as radios. ©WDP ©Hasbro Bradley Inc. and distributed thru PowerTronic by Nasta. Made in Hong Kong, and is 5H x 5-1/4W.

PLATE 173
WINNIE-THE-POOH. A Disney character with somewhat limited availability. ©WDP and distributed by Philgee International. Made in Hong Kong, and is 6H x 6-1/4W.

Courtesy BNZ

68

CHARACTER AND FIGURAL

This chapter feature characters and figures that are non-fictional. While a few, such as Elvis Presley and John Lennon, owe their fame to the media, they appear in this chapter because they were real celebrities rather than fictional characters.

Dogs, cats, and other animals also appear in this chapter, but these are a traditional representation of the animal rather than the almost "humanized" ones featured in the media section.

This chapter also introduces a new and ingenious device called the "Blabber Mouth." This clever invention uses a solenoid across the audio output circuit, connected to linkage that moves the figure's lips. If the radio is properly tuned, the lips appear to follow the sound coming through the loudspeaker. This system was developed and patented by Nasta Industries Inc. (through their "PowerTronic" division). The first radio to use this feature

was a plain white square unit with a pair of bright red lips on the front. This radio, called "Blabber Mouth," was introduced in 1985 (Plate 178). "Blabber Mouse," shown in Plate 179, was also introduced that year. Other units quickly followed: "Blabber Puppy" (Plate 181) was released in 1986, while "Money Talks" (Plate 211) — issued in 1987 — appears to end this series. Some other foreign-made units use a similar system, but don't honor Nasta's patent. They use the terms "Chatter" and "Talking" instead of "Blabber." A group of these radios is shown in Plate 180.

Another series of radios that appear throughout this chapter are the "Tune-A-Xxxx" series. These units are all made in Hong Kong and have no other markings on either the box or the radio. My personal favorite is the dog shown in Plate 188. This same series also includes the "Tune-A-Apple," shown in a later chapter.

The "dolls" shown in Plates 181-183 will have special interest for the adult male readers. The first two are mold variations, using different wigs. The last unit is almost identical in appearance, but is much smaller — only nine inches high.

This chapter also features a series of "memorial" radios honoring John Lennon, Elvis Presley, and John Wayne. These are all made on a "base radio" concept (with minor variations) and utilize a doll-like figure to portray the character. Since the doll is the critical part of the unit, and would be inexpensive to modify, perhaps other units exist.

Two other radios appear that are "commemorative" rather than "memorial." They were distributed by Radio Shack and honor the 200th anniversary of the United States and the Statue of Liberty re-conditioning project. Both units are in cast metal on a plastic base and exhibit excellent detail.

PLATE 174
ALLIGATOR RADIO. Fearsome looking Alligator, with the eyes used as windows to display the settings on the controls. Distributed by General Electric, although made in Hong Kong. It is 11-1/2L x 4W.

Courtesy BNZ

PLATE 175
ADAM AND EVE. While these units are very modern in style, they are anatomically correct and "mate" together. Marked "Trade Power" and made in Hong Kong. Each unit is 7-1/2H x 2-1/4W.

Courtesy LCB

PLATE 176
TUNE-A-BEAR. While he "Bears" a strong resemblance to Yogi, he is really a generic type, complete with bow tie. Made in Hong Kong and is 7H x 4W.

PLATE 177
BEEGEES. Sing-A-Long unit featuring the "Brothers Gibb." ©1977 Vanity Fair and licensed by the Image Factory, Hollywood, California. Made in Hong Kong and measures 8L x 7-1/2H.

Courtesy BNZ

PLATE 178
BLABBER MOUTH. This AM/FM unit was the first in the "Blabber" series. The bright red lips set off a pair of pearly white teeth that seem to follow the sounds coming out of the speaker. The term "Blabber" is copyright 1985 by Nasta. Made in Hong Kong, and is 5 inches square.

PLATE 179
BLABBER MOUSE. Second unit in the "Blabber" series. This darling little fellow comes complete with his slice of cheese. ©1985 Nasta Industries Inc. Distributed through their PowerTronic Division. (The same unit was also sold through Radio Shack.) Made in Hong Kong, and is 6-1/4W x6-3/4H.

PLATE 180
Group of three "Blabber" type radios, but use words like "Chatter" and "Talking" to avoid Nasta's Patent. The two blue units are indentical except for decal. They are made in Taiwan and are 5-1/2W x 5H. The center unit has no markings, and measures 4-3/4W x 4-3/4H.

PLATE 181
BLABBER PUPPY. Third unit in the "Blabber" series. This plastic bodied dog is covered with a flocking to resemble fur. ©1986 by Nasta Inc. Distributed through PowerTronic. Made in Hong Kong, and is 8H x 4W.

PLATE 182
TUNE-A-BULLDOG. Looks fierce enough to guard your house. Made in Hong Kong and measures 5-1/2H x 3-1/2W.

PLATE 183
JIMMY CARTER PEANUT. Clever design using the big toothed grin, the peanut shell, and the top hat associated with our former president. ©1977 Kong Wah Instrument Co. This Hong Kong unit is 7-1/2H x 3 Diameter.

PLATE 184
TUNE-A-CHICK. One of the smallest in the "Tune-A-(Animal)" series. Made in Hong Kong and measures 4-1/2H x 4W.

PLATE 185
CLOWN JUGGLING BALLS. Brightly colored unit, with the center ball having a light that flashes to the speech or music. Made in China, and is 9H x 6W.

PLATE 186
TUNE-A-COW. While she's a little darling, she's also cross-eyed. Made in Hong Kong and is 5-1/2H x 3-1/2W.

PLATE 187
TUNE-A-CAMEL. Nice unit, complete with
bell. 8L x 4-1/4W.

PLATE 188
TUNE-A-HOUND. This lovable little Pooch has a knit hat and
button eyes that move when you shake his head. Made in Hong
Kong and is 6H x 3-1/2W.

PLATE 189
TUNE-A-DUCK. While he has a sailor hat, I don't believe he is
supposed to represent Donald Duck. Made in Hong Kong and is
6H x 4 diameter.

PLATE 190
TUNE-A-FROG. This little guy, leaning on one arm, looks like he has lost his last friend. 6-3/4L x 4-1/2H.

PLATE 191
DOLL. This is not a childs doll, rather something for the mature male. Note her life-like hair, the pearl necklace, and the "see-through" nighty. The controls are via what I call "breastwork" tuning. (Modesty prevents more vivid descriptions.) No distributor information, but made in Hong Kong and is 12H x 8W.

PLATE 192
DOLL. This is made in the same mold as Plate 191, except the movable head is at a slightly different angle, and she has blonde hair. The "breastwork" controls are somewhat more visible.

Courtesy LCB

PLATE 193
DOLL. While she appears to another duplicate, this unit is much smaller and is only 9 inches high. A brunette version probably exists. Made in Hong Kong and is 9H x 8W.

Courtesy LCB

PLATE 194
TUNE-A-ELEPHANT. If blue Elephants are your thing, here's the one for you! Made in Hong Kong and is 5-1/4H x 4-1/4W.

PLATE 195
HAPPY FACE RADIO. This is a young childs radio that cannot be left on. It has a timer that only allows it to play for about five minutes. The eyes are the controls, while the nose is the push button to make it play. ©1987 Nasta Industries Inc. and distributed through PowerTronic. Made in Hong Kong and is 5-1/2H x 3-1/2W.

PLATE 196
Item A. DONT TOUCH THAT DIAL. An AM/FM unit showing a rather large arm and hand projecting out of the center.
Item B. INFINITY. Also AM/FM radio with man looking around radio and back to himself. Both distributed by Concept 2000. Made in Hong Kong, and both are 6H x 4W.

Courtesy LCB

PLATE 197
LADYBUG. This large unit is actually a radio/phonograph that plays 45 and 33 RPM records. The radio controls are via the "Eyes." Distributed through Radio Shack and made in Taiwan. It measures 15L x 9W.

PLATE 198
This is the unit described in Plate 197 with the lid open to show the phonograph and the controls.

PLATE 199
LADYBUG. Clever unit where the volume control opens the wings. Other color variations exist. British registration number 972006 and made in Hong Kong. It is 4-1/2L x 3-1/8W.

PLATE 200
MICHAEL JACKSON. This is an AM/FM unit with a paper decal of Michael on the front. It also features "slider" controls. ©1984 MJ Products Inc. Made in Hong Kong, and is 5-3/4H x 2-3/4W.

PLATE 201
JOHN LENNON (1940-1980). This is one of a series of "memorial" radios that appear in this chapter. This one features John Lennon. Elvis Presley and John Wayne also appear. All are made in Hong Kong and use a "base" radio concept. It is 10H x 5W.

PLATE 202
HELLO KITTY. Cute childs radio with a molded kitty on the front. Made in Taiwan and distributed by Samrio Co. Ltd. It is 5-1/2H x 3-1/2W.

Courtesy BNZ

PLATE 203
TUNE-A-LEO. One of the weakest renderings in this series. Made in Hong Kong and measures 5H x 4W.

PLATE 204
LADY IN HOOP SKIRT. This radio is made of ceramic, and manufactured by Radio Ceramics of California. Ceramic would have be considered an unusual material for a radio. Made in USA and is 10-1/2H x 9W.

Courtesy LCB

PLATE 205
LOVE IS . . . FOR US. Cartoon strip variation of Adam and Eve Theme. Nice detail and coloring. ©1973 Los Angeles Times and distributed by Sutton Associates Ltd. Made in Hong Kong, and is 7-1/2L x 4-5/8H.

Courtesy LCB

PLATE 206
LUCKY HORSE SHOE RADIO. Nicely detailed unit, complete with horse shoe. Another "base" radio, with visible thumbwheel controls. Made in Hong Kong, and is 5H x 4-1/2W.

Courtesy LCB

PLATE 207
MALE CHAUVINIST PIG. Why he has to be portrayed as a grouch, I don't know. Really cute unit with the curly tail acting as a volume control. The thumbwheel at top is for tuning. Made in Hong Kong, and is 6-1/4L x 3-1/2H.

PLATE 208
BOY WITH CANDLE. This figure is one of a series of four. It came in a box, (Pictured), with each side representing one of the characters. The box is marked "Marksons' Lovables Radio" on the front and "Solid State 7" on the back. The units are made in Hong Kong. This one is 8H on a red 4-1/4" Base.

PLATE 209
BOY WITH SUITCASE. This unit is pictured on the box mentioned in Plate 208. It happens to be on the side marked "Solid State 7" (Not shown). It is 7-3/4H and sets on a black base the same size as the one above.

Courtesy LCB

PLATE 210
BOY WITH WATER BUCKETS. Although I don't have the radio, this is one of the other sides of the box having pictures of the four characters. I would estimate that the measurments are close to the other two.

PLATE 211
MONEY TALKS. Another AM/FM version of the "Blabber" series distributed through Nasta. George's mouth follows the sounds coming out of the loudspeaker. ©1987 Nasta Industries and made in Hong Kong. It measures 8-1/4L x 3-1/2H.

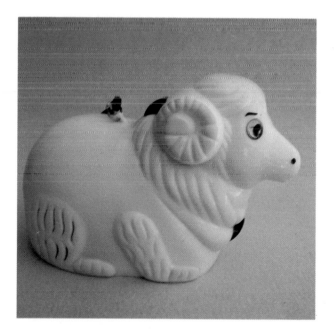

PLATE 212
BOY PLAYING INSTRUMENT (?) This is the final side of the box pictured in Plate 208. All of the figures appear to use the same base unit that actually contains the radio.

PLATE 213
TUNE-A-SHEEP. Typical unit in this series, with visible thumbwheel under chin. Made in Hong Kong, and has no other markings. Measures 6W x 4H.

PLATE 214
TUNE-A-MONKEY. This smug looking fellow is very unusual, as he has white fur! Looks like he knows all of the secrets of the universe. Made in Hong Kong and measures 6H x 3-1/2W.

PLATE 215
MONKEY HEAD. Cute little monkey with button eyes that move when you shake his head. Also pictured is the original box. British design number 964288, and distributed by International. Made in Hong Kong, and is aproximately 3-1/2 in diameter.

PLATE 216
MOUSE. Impish looking mouse mounted on a swivel base. It is made in Hong Kong and has no other markings. Measures 5-1/2H x 4 inches in diameter.

Courtesy BNZ

PLATE 217
TUNE-A-PIG. Also is a bank. Cute little pig with long eye-lashes. Made in Hong Kong and measures 4L x 4-1/2 in diameter.

PLATE 218
MY COUNTRY KID. This Chinese unit has no distributer
information, but is marked "Model LT-278R." It comes in
different colors. (Yellow and Blue versions shown here.) It is
6-1/2H x 5-1/2W.

PLATE 219
OWL. This early unit from Japan is typical of their work. Note
the jeweled eyes that are the controls. Plastic body with BRASS
Plated trim. It measures 6-1/2H x 4W.

PLATE 220
OWL. Another version of the old hooter. This one is all plastic and
the eyes are also the controls. Not as nicely detailed as the metal
version. Made in Hong Kong, and distributed by Stewart. 4-1/2H
x 3W.

PLATE 221
ELVIS PRESLEY (1935-1977). This is another "memorial" radio featuring Elvis in his famous "Thunderbird" costume. Same as details as Plate 201.

PLATE 222
PANDA BEAR. This little fellow seems to be smiling at you. The controls are via the knobs that make up the eyes. Made in Hong Kong, and distributed by Luxtone. 6H x 4-7/8W.

PLATE 223
PANDAS. Nice 2D rendering of the lovable Pandas. Made in Hong Kong, and distributed by Amico Inc. British design number 959481 and ©1982. It is 6W x 4-3/4W.

PLATE 224
ELVIS PRESLEY (1935-1977). Another version of Plate 221. Elvis is shown in a white costume in this unit. I have seen this same figure in one other costume, so possible other versions exist. No Distributer information. Made in Hong Kong, and is 10H x 5W.

Courtesy WMC

PLATE 225
RABBIT. No markings of any type. It is a very nice representation of a running rabbit in a 2D format. Made in Hong Kong, and is 7-3/8L x 6-1/2H.

Courtesy LCB

PLATE 226
TUNE-A-RABBIT. Nice detail, but is a typical representation of a classic rabbit and not Bugs Bunny. Made in Hong Kong and measures 7-1/2H x 3-1/2W.

PLATE 227
STATUE OF LIBERTY. This is a cast metal statue on a plastic base that contains the radio. It was sold through Radio Shack stores, where a portion of the purchase price went toward the statue reconditioning project. ©1986 by Tandy. Korean unit measures 11/34H x 3-1/2W.

PLATE 228
RADIO SHIRT. This Hong Kong unit is British design Number 100915, and is certainly unique. The controls are thumbwheels at the shoulders. No distributer information. It measures 4H x 5W.

Courtesy WMC

PLATE 229
SNORKS. Strange looking little fellow that looks like a squash with a handle. Authorized by Wallace Berrie and distributed through PowerTronic by Nasta. Hong Kong unit measures 5-1/2H x 5W.

PLATE 230
TURTLE. A nice Japanese unit made of metal and plastic. A similar Hong Kong unit in plastic also exists, but the turtle is not wearing a hat! It is 6L x 3H.

Courtesy WMC

PLATE 231
SPIRIT OF 1776. BiCentennial Commemrative marked 1776-1976. This unit was sold through Radio Shack stores to celebrate the 200th anniversery of the United States. Cast metal figures on a plastic base, similar to the Liberty unit in Plate 227. Made in Japan, and is 7-1/4H x 4-1/8W.

PLATE 232
TUNE-A-TIGER. Complete with bib overalls. Made in Hong Kong and is 5-1/2H x 4-1/4W.

PLATE 233
JOHN WAYNE "THE DUKE" (May 1907 - June 1979). Another in the "memorial" series, this time honoring John Wayne. The base treatment is slightly different. Made in Hong Kong, and same details as Plate 201.

PLATE 234
TUNE-A-WHALE. This unit has a movable tail and is somewhat unusual subject matter. Made in Hong Kong and measures 6L x 4-1/2H.

Robots, Rockets, and Outer Space

Robots have been objects of fascination for many years. One of the major attractions of the world fairs held in 1939 was a smoking, walking, talking robot called "Elecktro." He was accompanied by his robot dog, "Sparko." Although crude by today's standards, they represented the peak of technology for that period.

After World War II, science fiction movies such as "Forbidden Planet" and "The Day The Earth Stood Still" featured robots that were faithful servants. Another movie from that same era, "War of The Worlds," showed robots in a somewhat different light — that of an invader. Movies and television pictured a world where robots and space travel were firmly bonded.

The launching of the Russian satellite "Sputnik" in the 1950s finally crossed the line between fact and fiction — space travel was now reality. Those who watched the United States' efforts to challenge the Russians will never forget the day Neil Armstrong and Edwin Aldrin Jr. landed on the moon and stated, "One small step for man, one giant leap for mankind."

The science-fiction masterpiece, "Star Wars," released in the 1970s, generated another rage of "robot mania." The two lovable robots in that movie, R2D2 and C3PO, have been made into about every device you can imagine, but for some strange reason, never a transistor radio.

The peak period of robot collecting was from the mid-1950s through the 1960s. Many fine robots were made as battery-operated tin toys, and again our Japanese friends dominated this field. If you want to see outrageous prices for collectibles, this is the hobby to embrace. It is not unusual to see individual robots sell for several thousands of dollars! While these prices tend to limit the field to adults, children are also a ready market.

While the "true" robot collector scorns plastic with the same fervor as an "olde" toy buff, they are facing reality and recognize that the days of affordable robots in metal are a thing of the past. These collectors are now branching out into plastic robots, and the units pictured in this chapter — being both a robot and a radio — make this conversion somewhat less painful.

As with most items, those that have the best detail and exhibit the best craftsmanship will command the highest price. A good example is the "Star Command" series made by Calfax Inc. This series includes two "Starriods" (Plates 243 & 244), a robot mascot (Plate 245), and three space ships that are AM radios (Plates 246 Through 248). Two other units in the series are robots that have an AM radio and feature a Sing-a-Long mike (Plate 252 for

the only one of the two that I have). All of these units have excellent detail and look very futuristic. This type of radio will always have appeal to collectors.

Also included in this section are Transformers, Gobots, and other futuristic-looking figures from the media. It also includes some miscellaneous objects such as Armillary Spheres and space capsules. While "Battlestar Galactica" is represented I have not been able to find a single radio based on ANY character from the "Star Wars" series, although they are well-represented by a multitude of other objects.

Robots, rockets, and outer space items are considered "crossover" items and are usually priced accordingly.

PLATE 235
ARMILLARY SPHERE. This ancient astronomy device was used to plot the stars. Made in Japan, and distributed through Heritage. The base unit has a presentation plate on it. 6-1/2W x 8H.

PLATE 236
CREATURE I. Nice friendly looking robot who is a digital clock as well as an AM/FM radio. Distributed by Timco, and is 5-3/4H x 6W. Made in Hong Kong.

PLATE 237
CYLON WARRIOR. Based on the Television series "Battlestar Galactica." ©1978 Universal Studios and distributed through Vanity Fair Industries Inc. ©1979. The face plate flashes with the sounds. Made in Hong Kong, and is 5-3/4H x 4W.

PLATE 238
RADIOBOT. This radio is distributed through Radio Shack stores. Made in Hong Kong, and is 8H x 4W.

PLATE 239
TIMES SPUTNIK. An obvious reference to the Russian satellite. This Japanese unit measures 5H x 4 inches in diameter.

Courtesy BNZ

PLATE 240
MOONSHIP. Same radio as Plate 239, but marked "Moonship." Perhaps the makers felt the term "Sputnik" wouldn't sell well in the USA (?) Same details as Plate 239.

PLATE 241
TRANSFORMERS. Autobot Freedom Fighter. ©1984 by Hasbro Industries Inc. and distributed by Nasta Inc. Made in Hong Kong, and is 4-1/4H x 4W.

PLATE 242
TRANSFORMERS. This unit has a large window just above his nose that flashes to the sounds. ©1985 Hasbro Inc. and distributed by Nasta Inc. Made in Hong Kong, and is 7-1/2H x 5W.

PLATE 243
STARROID IM1. (I AM ONE) This is the leader of the Star
Command. This series is distributed by CalFax, and is ©1977. All
Robots in this series have flashing lamp eyes that follow the sound.
Made in Hong Kong, and is 8H x 5-1/2W.

PLATE 244
STARROID IRI2. (I ARE ONE TOO). The second unit of the
Star Command. distributed by CalFax and ©1977. Made in Hong
Kong, and is 8-1/4H x 5W.

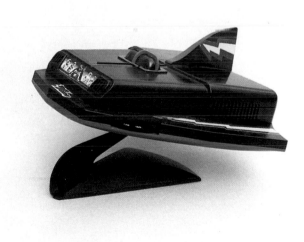

PLATE 245
ROBARK. The mascot of the Star Command. He is ©1979 and
same distributor as above. Made in Hong Kong, and is 6-3/4H x
5-3/4W.

PLATE 246
DARK INVADER. Spaceship that is part of the Star Command.
(Another version of this mold appears in Plate 247.) ©1977. Made
in Hong Kong, and is 7L x 7-1/2W.

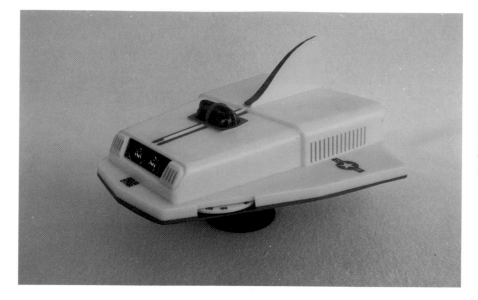

PLATE 247
AQUASTAR. This is another spaceship in the Star Command. (Same mold as Plate 246.) It also features the flashing lamps in the cockpit. ©1977 CalFax. Made in Hong Kong, and is 7L x 7-1/2W.

Courtesy BNZ

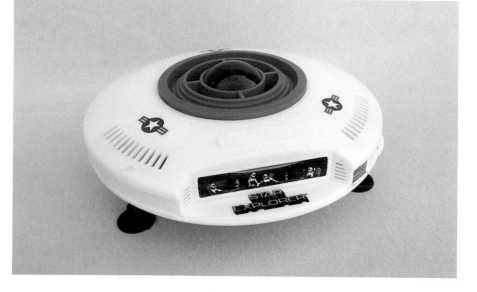

PLATE 248
STAR EXPLORER. Assumed to be the "Mother" ship of the Star Command. This also uses flashing lamps in the dome. (There is also a Star Explorer II, made in black.) ©1977 by CalFax. Made in Hong Kong, and is 7 inches in diameter and 3H.

Courtesy BSG

PLATE 249
SPACE SHUTTLE COLUMBIA. This radio was distributed by Tandy through the Radio Shack Stores. It is made in Hong Kong and measures 6-1/2H x 10-1/4L.

PLATE 250
ROBOTIC RADIO. This is another transformer unit that folds into a suitcase shape. ©1984 Universal Studios and distributed through Royal Condor ©1984. Made in Hong Kong, and is 6-5/8H x 7W.

PLATE 251
ROBO-AM1. Another unit featuring the flashing eyes. Made in Hong Kong and distributed by Westminster. It is 8-3/4H x 6W.

PLATE 252
STARROID IR4U. (I ARE FOR YOU). This unit is a sing-a-long, with the flashing eyes. Distributed by CalFax and ©1977. Made in Hong Kong, and is 9H x 5-1/2H. (Missing microphone.) There is one more robot in this series called IM4U. (Not Pictured.)

PLATE 253
ROBOT WITH CLOCK. This fierce looking fellow is Made in Hong Kong and distributed through "Equity." It is 9-1/4H x 6W.

Courtesy LCB

PLATE 254
ROBOCHANGE. (ROBOT MODE.) Robot made from the car shown in Plate 26. ©1985 TaiFong. Made in Taiwan and is 9-1/4H x 6D.

PLATE 255
RADIO GOBOT. (ROBOT MODE.) This is the robot made from the car pictured in Plate 25. These seem to lack the detail of the other robots, and look just like a robot made from a car. ©1985 Tonka Corp. Distributed by Playtime Products. Made in Hong Kong, and is 8-1/2H x 4-1/2D.

PLATE 256
MR. D. J. This AM/FM unit is the final word on Robots! His eyes blink, his mouth moves, his arms swing, and his body rocks from side to side — all in sync with the sound! Distributed by Tomy (Model 5420), and made in Hong Kong and is 6-1/2H x 5W.

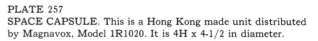

PLATE 257
SPACE CAPSULE. This is a Hong Kong made unit distributed by Magnavox, Model 1R1020. It is 4H x 4-1/2 in diameter.

PLATE 258
FLYING SAUCER. This spaceship is marked "Realtone" and is a Hong Kong unit measuring 3-3/4 in diameter and 2-3/8H.

PLATE 259
POWER BOX RADIO. Transformer unit similar to Plate 250, but marked "Best Join - ©1985 USA" on the rear of the unit. Made in Taiwan and measures 8H x 8W as shown.

NOTES

Advertising & Product-Shaped

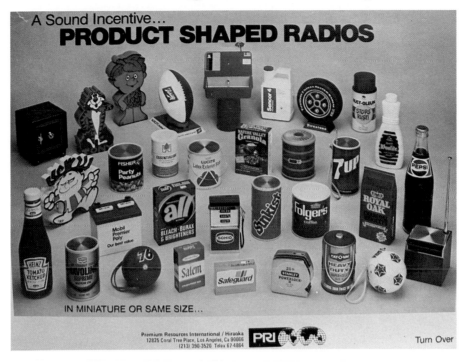

Fig. A - Some "product-shaped" radios. Note parts washing sink. (At top center.) The round object next to the 7-Up can is an air conditioner unit designed for the Carrier Corp.

One of the more interesting fields of novelty radio collecting is "product-shaped" radios. This family includes a wide range of items from the popular "can" radio to such diverse things as a tape measure, a ham can, and a tire (plus a myriad of other items).

One of the larger distributors of these radios is Premium Resources International (also Known as PRI/HKA or PRI/Hiraoka). This is a sales promotion company, associated with a trading company. Their radios are manufactured in Hong Kong.

Thanks to Jerry Koenig of PRI, I was able to obtain a series of catalogs and articles covering their products over a period of years. According to Jerry, most of their radios are "pre-sold" and shipped directly to the company ordering them. They usually are a premium or an advertising campaign item to encourage sale of a product.

PRI will work with a company to adapt one of their existing molds to a prod-

uct, or will design a new one to the customer's specifications. They will also help develop promotional material for the company buying their radios.

Many of these radios were never intended for release to the general public, but were used as an "in-company" award for sales effort, attendance, or other noteworthy accomplishments. Others, like the beverage cans, batteries, and the tape measure, were used as an incentive to purchase a product — usually awarded to the buyer with proof-of-purchase and a few dollars to cover postage and handling.

The use of mold variations was covered in the "Travel" section, but it is in this one that the technique appears almost to excess. Once a mold is made it can be adapted to many products, even for different companies, usually by just changing the label. Thus the paint can for Lucite becomes Valspar Varnish, and by simply removing the handle, the mold is then used for Del Monte Pineapple, Fisher Peanuts, and Folgers Coffee. To add further challenge to your collecting efforts, a few units feature different products from the same company on opposing sides of the radio (Borax Soap/Borax Bleach in Plates 260 & 261 and the Carnation Chocolate can shown in the next chapter are good examples). They complicate displaying the unit because you always wonder which side to place out. (This isn't a problem if you have TWO of the units.)

The "can" units really become only label variations, and the radios made in these molds are among the most common ones you will find. (More than 50 variations appear in this book, and that is just a sample.) Two basic can molds are used with minor differences among manufac-turers: One, used for beverage cans, is very close to actual size (4-3/4H x 2-5/8 Dia.) and about every beverage you can imagine is represented. The other mold is used for soups, motor oils, and a few miscellaneous items. When this mold is used for soups, it, too, is close to actual size (4H x 2-5/8 Diameter,) but when used for motor oil it has the wrong aspect ratio and loses accuracy. In actual practice, the oil can is taller than either mold and has a greater diameter (5-1/2H x 4 Diameter).

Oil is now sold in a plastic container resembling the "Sencor 4" unit shown in Fig A. Perhaps a new group of product-shaped units will be available when the oil companies want to update their advertising.

PRI also markets radios under the

Fig B. Some of the many "can" units. These are one of the more common product-shaped radios around. (Many types and manufacturers. The one on the lower right is an AM/FM unit.)

name "Brandsounds." The owners of certain products have entered into licensing agreements with PRI to allow direct sale of their radios to the public. Brandsounds packages these radios into multiple displays and sells them through various outlets, such as drug stores and super markets. These radios will have a much greater exposure than those that are limited to distribution through the owner of the product. This is why you will find many more "can" radios than the tape measure or oil filters.

While PRI uses the term "product-shaped," I prefer to classify these units as advertising items. Many novelty radios are product shaped — cars, boats, and even pianos are products in the strictest sense of the word.

This chapter will only feature radios that have a brand name (or a recognized trademark) AND the unit could be used to encourage sale of a product. It is also reserved for non-food and drink items. (Food and drink items appear in the next chapter.) A few food or drink logos, such as Tony-the-Tiger and Punchy may appear, but not the actual product.

While several interesting radios appear, the two spark-plug units shown in Plates 314 and 315 represent one of the rare cases in which a Hong Kong unit is actually worth more than the Japanese one. In this case, the Hong Kong unit is AM/FM and line powered, while the Japanese one is AM only and uses a battery.

The use of a novelty radio for an award is demonstrated by the "Big A -

Fig C. - More "product-shaped" radios. Note Ham Can, Oil Filter and Tire.

Joint Management Meeting" shown in Plate 264. This car battery was probably passed out to everyone who attended this meeting, and ended up by the dozens in desk drawers or at home on the dresser. (Pacific Telephone, my employer, used digital clocks for their meetings.)

The Raid unit in Plate 303 is another example of an "in-house" unit. According to the person I purchased it from they are not distributed to the general public, but used to reward attendance, or for other noteworthy accomplishments. This particular individual had gathered several over his career and was willing to part with one.

If an award was presented for the most appealing unit in this section, my vote would be for the "Helping Hand" shown in Plate 284. This unit always attracts attention when I display my collection. There is just something about the style — and the cute little grin — that draws your attention to him.

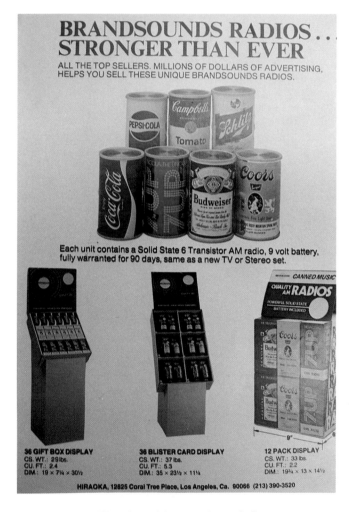

Fig D. - A typical "Brandsounds" display for use in Supermarkets and other stores.

Fig E. - Some other products licensed to "Brandsounds."

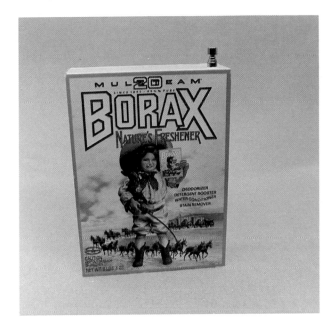

PLATE 260
BORAX SOAP. The famous "20 Mule Team" in a two-sided radio. (Other side in Plate 261.) This is a PRI mold and measures 4-3/4H x 3-1/2W.

Courtesy BNZ

PLATE 261
BORATEEM BLEACH. This is the other side of the radio shown in Plate 260.

Courtesy BNZ

PLATE 262
BON AMI POLISHING CLEANSER. "Recommended by Corning." Also shows the famous baby chick, with the motto: "Hasn't scratched yet." Made in Hong Kong and is the same size as a beverage can.

Courtesy LCB

PLATE 263
CARRIER HIGH EFFICIENCY. This is an air conditioner unit made by PRI for the Carrier Company. It is made in Hong Kong and measures 4H x 3-1/2 in diameter.

Courtesy LCB

PLATE 264
CAR BATTERIES. The "Big A ... Joint Management Meeting" is an example of using a novelty radio as an award. This was probably passed out at this meeting in April of 1980. Other unit is "Trop Artic - Maintenance Free" Both units:

Courtesy LCB

PLATE 265
CAR BATTERIES. Old style with terminals out the top. These terminals are the controls of the radio. These two happen to be ATLAS and MOBIL, but many versions exist. They are PRI units, made in Hong Kong, and are 4W x 3-7/8H.

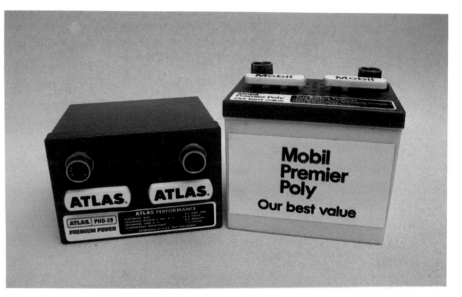

PLATE 266
BATTERIES. Texaco "Super Chief" and Co-op "Forget It." Same dimensions as Plate 265.

PLATE 267
CAR BATTERY. New style, with terminals out the side. This one is a "Delco Freedom Battery," but many others use this mold. It uses the terminals for the controls. Made in Hong Kong, and is 4-1/4H x 4W.

PLATE 268
BATTERY. This "D" cell is marked "Radio Shack, New Formula, Steel Clad, 1.5 Volt" This unit is obviously distributed through Radio Shack stores. It is made in Korea, and is 4-3/4H x 2-1/4 in Diameter.

PLATE 269
BATTERY. Ray-O-Vac Super Cell. Also marked "Heavy Duty." This unit is also marked ESB Inc. Ray-O-Vac Division - ESB Canada Ltd. It is made in Hong Kong, and is 4-1/4H x 2-3/4 in Diameter.

PLATE 270
CARTER'S PEANUTS STYLE RADIO. An obvious play on Jimmy Carter. This is marked "Model PR-1" and is distributed by Windsor Industries. It is made in Hong Kong, and is 3-3/4H x 3-3/8 in diameter.

PLATE 271
BOX RADIOS. This mold is used for many radios. NESTLE SOUP and CHAMPION SPARK PLUGS here, but many other units use this PRI mold. Made in Hong Kong, and are 4-3/4H x 3-1/2W.

PLATE 272
CHARLIE THE TUNA. Charlie is shown molded on a bicycle radio designed to clamp on the handlebars. The words "Sorry Charlie" refer to his being rejected for not meeting Starkist standards for their tuna. ©1973 by Starkist Foods. Hong Kong unit measuring 3-1/4W x 5-1/4H.

Courtesy BNZ

PLATE 273
CHARLIE THE TUNA. The more classic version of the rejected tuna. This radio is ©1970 by Starkist Foods. (The base is not original on this unit.) Made in Hong Kong, and is 5-1/2H x 3-1/2W.

PLATE 274
CIGARETTE PACKAGES. Two different packs are shown here. The "soft" pack is 4-1/8H x 2-1/2W. The "flip top," showing the controls, is 3-3/4H x 2-1/4W.

Courtesy LCB

PLATE 275
KENT CIGARETTES. This large unit is AM/SW and is made in Japan. It is 13-1/2H x 8-3/4W.

PLATE 276
KENT CIGARETTES. This is the "flip top" pack, shown open to expose the controls. Made in Hong Kong, and is 3-1/4H x 2-1/4W.

PLATE 277
DERMOPLAST. Anesthetic skin spray. The spray is distributed through Ayerst Laboratories, Inc. - but no other markings on the radio, except "Made in Hong Kong." It is 6-3/4H x 2-1/4 in diameter.

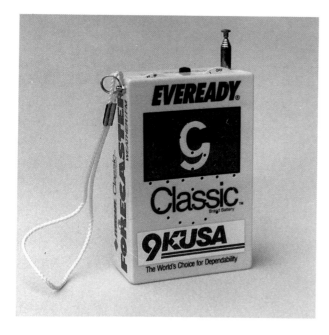

PLATE 278
EVEREADY CLASSIC. "The Worlds Choice For Dependability." This radio is AM and also tunes weather channels. It also features the call sign "9KUSA." Made in China, and is 4-1/2H x 3W.

Courtesy LCB

PLATE 279
BEST BUY. Use of a "can" radio for an advertising item. General Electric logo with "Our No. 1 Goal - Quality - Service - Value." Made in Hong Kong, and is standard size.

Courtesy BNZ

PLATE 280
FAULTLESS SPRAY STARCH. "Look and Feel Fresh All Day."
Made in Hong Kong, and is 6-1/2H x 2-5/8 in diameter.

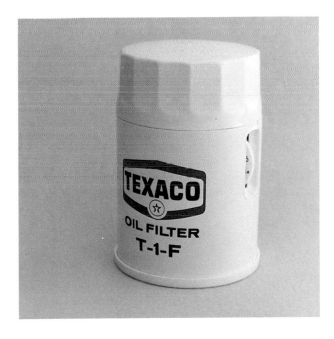

PLATE 281
FILTER. Texaco unit marked "T 1 F," but other types and sizes
available. PRI mold and measures 4-1/2H x 2-3/4 in diameter.

Courtesy LCB

PLATE 282
GETTY GASOLINE. A 1950s style gas pump. Another group of
these units appear in Plate 283. PRI mold, and is 4-1/2H x 2-
1/4W.

Courtesy LCB

PLATE 283
GAS PUMPS (1950s Style). These use the windows of the pump for the tuning indicators. Many different versions exist. (The Sinclair one is very early.) They measure 4-1/2H x 2-1/2W.

PLATE 284
HAMBURGER HELPER. This is the "Helping Hand," and is a trade mark for General Mills. At this writing, the current offer is this "Hand" with a clock in the palm. This unit uses the red nose for the volume control. Made in Hong Kong, and is 6-1/2H x 5-3/4W.

PLATE 285
HOELON 3EC HERBICIDE. Also marked "The Practical hoe for grass control." Made in Hong Kong, and is 4-1/4H x 3-1/4 in diameter.

Courtesy LCB

PLATE 286
GAS PUMPS. Another pair of 1950s style gas pumps. These two are by Texaco, featuring "Lead-Free Gasohol," and "Fire Chief." Same size as Plate 283.

Courtesy LCB

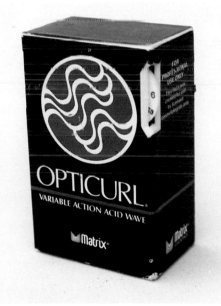

PLATE 287
KOSI LITE. No, not a funny beer, but a Radio Station in the Denver area that plays "Lite" music. It is an AM/FM unit the same size as a beverage can.

Courtesy LCB

PLATE 288
OPTICURL. "Variable Action Acid Wave," and sold by Matrix. No other markings on unit. Made in Hong Kong, and is 4-1/4H x 2-5/8W.

Courtesy LCB

PLATE 289
OIL CANS. These are among the most common of the novelty radios. Havoline "Super Premium," and Mobil "Delvac 1200." Each 4H x 2-5/8 in diameter. This is the standard size for these cans.

Courtesy LCB

PLATE 290
OIL CANS. Havoline "Super Premium" and Shell "Super X." Standard size.

PLATE 291
OIL CANS. Mobil "Super" and Chevron "Custom Fuel Saving." Standard Size.

Courtesy LCB

PLATE 292
OIL CANS. Mobil "1" and Chevron "Custom Motor Oil." Standard Size.

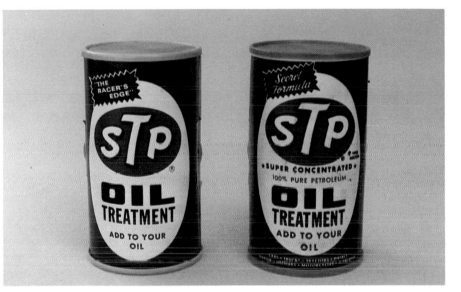

PLATE 293
STP OIL TREATMENT. Two different versions of the same product. Left is "The Racers Edge" while the one on the right is the "Secret Formula." Standard size.

Courtesy LCB

PLATE 294
PUNCHY. This is the Logo of Hawaiian Punch. It is a PRI product, made in Hong Kong, and has British design number of 971396. It is distributed by R. J. Reynolds Tobacco Co. via Specialty Advertising Products (Code No. 14250), Winston-Salem, N. C. It measures 7W x 6H.

PLATE 295
LOGO. This ball shaped unit is the logo for the Union Oil Company and bears their famous "76" markings. It is reminiscent of the styrofoam balls the station attendants used to place on your car antenna back in the 1960s. Made in Hong Kong and is 3-1/2 inches in diameter.

Courtesy LCB

PLATE 296
LOGO. This logo is the trade mark of the Shell Oil Company. It is a company sponsored item, and somewhat rare. Made in Hong Kong and measures 4-3/4H x 5-1/4W.

Courtesy LCB

PLATE 297
OIL CAN. "Stihl Chain Saw Engine Oil." One of the rarer units in the oil can series.

Courtesy BNZ

PLATE 298
NEWS. A "shirt pocket" unit in the shape of a vending machine, selling the "Rocky Mountain News." The radio has no other marks and measures 4H x 2-1/4W.

Courtesy LCB

PLATE 299
PAINT CANS. BOYSEN and DIAMOND VOGEL pictured here, but many others exist. These are probably PRI products. 3-3/4H x 3-1/4 in diameter.

Courtesy LCB

PLATE 300A
PAINT CAN. "Cook Paints and Coatings." Another in the paint can series. Made in Hong Kong, and is 3-3/4H x 3-1/4 diameter.

PLATE 300B
PAINT CAN. "Pratt & Lambert AquaRoyal Latex House & Trim Finish." This can contains "White" paint. Made in Hong Kong, and is 3-5/8H x 3-1/4 in diameter.

Courtesy BNZ

PLATE 301
PLANTERS PEANUTS. Mr. Peanut is a long standing trade mark, and is the property of Planter Peanut Company (No Date). It is a PRI mold, made in Hong Kong, and is 10-1/4H x 5W.

PLATE 302
PILLSBURY DOUGHBOY. A trade mark of the Pillsbury Company and ©1985. This is one of the few "walkman" type radios in this book. Made in Hong Kong, and is 6-1/2H x 3W.

PLATE 303
RAID. This digital clock and AM/FM radio was not offered to the general public, but limited to employee's of the company. It is marked "©S. C. Johnson & Son Inc.," and made in Hong Kong. It is 7W x 7H.

PLATE 304
RUST-OLEUM SPRAY PAINT. "Beautifies As It Protects" and
"Stops Rust." Metal primer unit made in Hong Kong. It is 6-1/4H
x 2-1/8D.

Courtesy LCB

PLATE 305
RAISIN MAN. This little bit of work took the nation by storm.
This AM/FM unit has posable arms. ©1988 by CalRab and
licensed by Applause. The radio is distributed by Nasta Industries
Inc. ©1988. Made in China and is 7H x 6-3/4W.

PLATE 306
SAFE. This unit is complete with wheels and a bear on the front.
It advertises MSI Bank Of Hilldale, and is probably made in Hong
Kong (No Marks). It is 4 x 4 x 5H.

Courtesy BNZ

PLATE 307
RADIO. This AM/FM unit was used as an advertising campaign with the theme of "Radio is Red Hot!" This radio frequently has a local station decal on the front. Isis model 20. Made in Hong Kong, and is 10L x 3-1/2H.

PLATE 308
RADIO. Same unit as above, but made for AVON and called Isis Model 20-1. Another red unit made by Isis is the word "CLOCK," having a small clock in the letter "O" of the word.

PLATE 309
RADIO SHACK. Logo in bright red. The openings in the "R" and in the "O" of the word "Radio," are used for the indicators on the radio controls. Made in Hong Kong, and is 6-3/4L x 4H.

PLATE 310
A pair of "shirt pocket" radios advertising local radio stations. Many different types exist, as they are ideal promotional units for a radio station.

Courtesy LCB

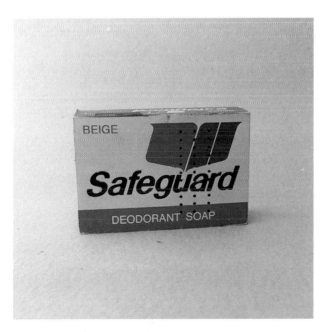

PLATE 311
SAFEGUARD. This is another of the "box" shaped radios. It is marked "Deodorant Soap," and the word "Beige" also appears on the front. Made in Hong Kong, and is 3-1/4L x 2-1/2H.

Courtesy BNZ

PLATE 312
SCHLITZ FOOTBALL. While it could also be placed under "Sports," or even under "Food & Drink," the style, and the "SCHLITZ makes it great!" tells me it's an advertising item. This is a PRI mold, made in Hong Kong, and is 7W x 8H.

Courtesy WMC

PLATE 313
CHAMPION SPARK PLUG. This is an AM/FM unit manufactured by PRI. It uses the high voltage connection for the volume control, and the base threads are for tuning. Made in Hong Kong, and is 9-1/2L x 2-3/4 in diameter.

PLATE 314
CHAMPION SPARK PLUG. This is an early Japanese unit that is AM only. The volume control is via the high voltage electrode, and it is tuned by turning the top insulator. It is 14-1/2H x 8-1/2D.

PLATE 315
CHAMPION SPARK PLUG. This is an AM/FM unit that is line powered. The logo on the base is also in colors, where the previous one is buffed metal. Same size as Plate 314. Made in Hong Kong.

PLATE 316
STANLEY TAPE MEASURE. This life size unit is very collectible. The controls are via thumbwheels that are almost flush with the top. PRI mold and made in Hong Kong. It is 3-1/8W x 3H.

PLATE 317
TONY THE TIGER. This smiling fellow is the Trade Mark of the Kellog Co, and is ©1980. It also has a British Registration Number of 993394 on the back. Made in Hong Kong, by PRI and is 7H x 4W.

PLATE 318
STEEL BELTED RADIAL TIRE. This PRI mold is used for many tires. This one is "Gulf Radial Steel 78" and the controls are the lug nuts in the center. Made in Hong Kong, and is 5 inches in diameter.

PLATE 319
TIRE. "Winston Winner." Slightly different version, having a carrying strap with the controls at the rear of the unit. Made in Hong Kong and is 5 inches in diameter.

Courtesy LCB

PLATE 320
LITTLE SPROUT. "Little Sprout" is the "Green Giants" son. They live in "Green Valley" and are Trade Marks of the Pillsbury Company. This is a PRI mold, made in Hong Kong, and measures 8-1/4H x 4-3/8W.

PLATE 321
WOOLITE. Marked "For All Fine Washables." Also allows "Cold Water Wash." PRI unit made in Hong Kong. 7-3/4H x 2-3/4W.

PLATE 322
ZANY. A perfume sold by AVON. This radio is made in Hong Kong, and is 4-3/4H x 5W.

Courtesy LCB

Food and Drink

With the exception of a few generic items, such as the hot dog, hamburger, apple etc., this chapter should be considered a continuation of the previous "product-shaped" section.

This chapter exist for two reasons: To place all of the food and drink items in one chapter, allowing quick access to these items, and to treat this section differently than the others in this book. Since a substantial portion of this chapter consists of beverage cans, shot in pairs, those captions will be brief. Most of the cans were purchased through the secondary market and providing distributor credit would be almost impossible. The trade-mark is evident, and since the cans are all close to actual size, dimensions are not necessary. Although a few of the more unusual units did warrant a comment, most of the captions will omit any other details as the pictures alone provide adequate information

to identify the unit.

The Pepsi soda fountain dispenser, shown in Plate 361, may have the distinction of being the most expensive radio in this book. The January 1989 issue of the "Antique Radio Classified" reports an auction that was held on November 19, 1988, in Peabody, Massachusetts, where a duplicate of this unit was sold for $1300! Hopefully, owners of this unit will realize that auction prices are among the least reliable ones in the collecting marketplace. Two bidders can work in collusion and bid the price to any level they feel appropriate. Even without collusion, auctions tend to create a "bidding fever," in which a buyer will be caught up in the emotion of bid and end up paying a price far above what they would normally pay. The real price of this unit (and all others in this book), will have to be determined over a period of time - not just a price based on a single sale.

Other note-worthy radios in this section are the "billboards" shown in Plates 335-337. While these show Pepsi and Coca-Cola units, the design of the radio — with the removable face plates — leads me to believe that other units may exist.

The most unique units in this chapter may be the "Cold Drinks With Ice" vending machine shown in Plate 385, and the very rough Coca-Cola vending machine shown in Plate 393 — both distributed by Westinghouse circa 1963. From the outside, they look just like any other vending machine radio. When opened, it is interesting to note that these units are nothing but a vending machine case containing a Westinghouse Model H928 "shirt pocket" radio. Careful inspection of Plate 394, the inside view of the Coca Cola unit, will reveal the small metal clamps and washers holding the pocket radio in place. The red lever switches the radio from AM to FM.

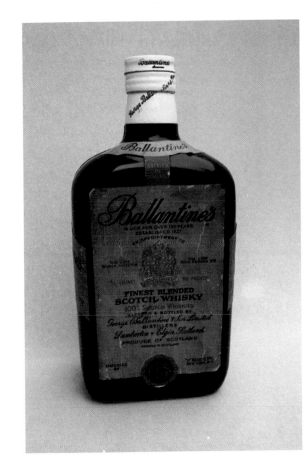

PLATE 323
BALLANTINE'S. Also marked "Finest Blended Scotch Whiskey." Distributed by Caltrade Manufacturing (No Date). Controls via split cap. Made in Hong Kong, and is 8-3/4H x 4-1/4W.

PLATE 324
BUDWEISER BEER BOTTLE. "King of Beers." Used with permission of Anhauser Busch. Made in Hong Kong, and is 9-1/2H x 2-1/4 in diameter.

PLATE 325
COKE BOTTLE. This actual sized unit is made in Hong Kong, but authorized by the Coca-Cola company. The volume control is via the top section, and the tuning is done by a movable "base" plate. Nice unit, but quite common.

PLATE 326
CAMUS GRAND V.S.O.P. Also marked "LaGrande Marque Cognac." The controls are via the "seals" on the front of the bottle. Made in Hong Kong, and is 10-3/4H x 3-3/4 in diameter.

PLATE 327
HERSHEY'S SYRUP. Also marked "GENUINE CHOCOLATE FLAVOR." No other markings. Made in Hong Kong, and is 8-1/4H x 4-1/4W.

PLATE 328
HEINZ TOMATO KETCHUP. Life sized unit that is somewhat rare. Hong Kong unit using same control techniques as Plate 325.

PLATE 329
FLEISCHMANN'S GIN. Unusual radio as it uses clear plastic and the actual radio is visible. Distributed by Prestige, and is 11H x 3-3/4W.

Courtesy LCB

PLATE 330
MALIBU. AM/FM unit marked "Caribbean Rum Top Grain Neutral Spirits . . . " It is distributed by Diamond Promotion Group, and made in China. Measures 11H x 3-1/4 in diameter.

PLATE 331
PEPSI. This is made in a PRI mold, but was also sold through Radio Shack stores. Same control technique as unit in Plate 325. Made in Hong Kong, and is 9-5/8H x 2-3/4 in diameter.

PLATE 332
PIPER BRUT CHAMPAGNE. Also marked, "Piper-Heidsieck" and the champagne is vintage 1961. Distributed by Raleigh Electronics and is Patent Number 204318. Controls via split cap. It is a Japanese unit measuring 10-1/2H x 2-3/4 in diameter.

Courtesy LCB

PLATE 333
TEACHER'S HIGHLAND CREAM. "Bottled In Scotland" This Japanese unit is distributed by Industrial Contacts Ltd. and is marked "Pat #68060 & 23926." The tuning and volume is via the split cap. It is 12H x 3 in diameter.

Courtesy LCB

PLATE 334
YAGO SANT GRIA. Also marked "A MONSIEUR HENRI SELECTION" and "Fine Spanish Wine and Citrus Juices Imported by Monsieur Henri Wines Ltd. White Plains NY 10604." The volume control is the cap and the tuning is via the control on the bottom. Made in Hong Kong, and is 10-1/2H x 2-1/4 square.

PLATE 335
BILLBOARD RADIO. This is an AM/FM unit distributed by IsIs (Model 1845). This one advertises "Enjoy Coca-Cola" and features the "wave." Made in Hong Kong, and is 12L x 4H.

PLATE 336
BILLBOARD RADIO. Same unit as Plate 335, but this time advertising "Pepsi."

PLATE 337
BILLBOARD RADIOS. This is the back view of both radios. The Coke unit uses the same message on the rear, while the Pepsi one adds "Catch That Pepsi Spirit."

PLATE 338
LEFT. BUDWEISER BEER. "King-of-Beers" in an AM only version.
RIGHT. SEAGRAM'S COOLER. Very glossy label that reads " The naturally flavored citrus and wine beverage."

PLATE 339
LEFT. CARNATION. This is a 2 sided can, showing "Hot Cocoa Mix" on this side. (Courtesy BNZ)
RIGHT. SEVEN-UP. This is an older style label where the words are actually spelled out. Darker green than newer versions.

PLATE 340
LEFT. CARNATION. This is the other side of the can shown in Plate 339. "Sugar Free" mix on this side. (Courtesy BNZ)
RIGHT. SUNKIST. Also states "Good Vibrations" on a bright orange background.

PLATE 341
LEFT. CERVEZA TECATE BEER. Brewed in Tecate, Baja California - a small border town near San Diego. Somewhat limited distribution.
RIGHT. SCHLITZ BEER. "The beer that made Milwaukee famous."

PLATE 342
LEFT. COORS "Americas fine light beer" AM/FM unit marked in 12 fld oz size.
RIGHT. BUDWEISER BEER "King Of Beers" in AM/FM unit. Also marked with 355ml.

PLATE 343
LEFT. COORS LIGHT In aluminum color can.
RIGHT. HAMM'S BEER. "Americas classic premium beer, born in the land of sky blue waters."

Courtesy LCB

PLATE 344
LEFT COCA COLA USA. Shows olympic rings around the bottom. Possibly for Los Angeles Olympics?
RIGHT. FRESCA. Enjoy sugar free! also "A unique blend of citrus flavors."

PLATE 345
LEFT. COKE. This unit has Japanese markings on it, and is a very early radio. The ends look like aluminum, and the radio features excellent workmanship.
RIGHT. MELLO YELLOW. Marked "A product of the Coca Cola Company" and Citrus flavors. 12 fl. oz. (355 ml).

PLATE 346
LEFT. COCA COLA. Enjoy Coca Cola with the "wave" symbol on side. (This is a very Glossy label.)
RIGHT. RC ROYAL CROWN COLA. Old style label.

PLATE 347
LEFT. COCA COLA. "Enjoy Coca Cola"
With the "wave," but in a dull paper label and
includes "12 FL. OZ."
RIGHT. DADS ROOT BEER. Marked 12 fl.
oz. and (355 ml).

PLATE 348
MUG ROOT BEER. Two different versions
of the same beverage.

Courtesy LCB

PLATE 349
LEFT. BARRELHEAD ROOT BEER.
Draft Root Beer in a 12 oz. can.
RIGHT. PEPSI. Modern logo in AM only
version.

PLATE 350
LEFT. PEPSI LIGHT. "Sugar Free Cola &
Lemon . . . "
RIGHT. COORS. "Brewed with Pure Rocky
Mountain Spring Water" in an AM version.

PLATE 351
PEPSI. Block letter style logo on a AM/FM
unit.
RIGHT. 7-UP. Red dot version of label. Also
marked 354 ml.

PLATE 352
This is the reverse sides of the two can units
shown in Plate 351. Note that the logo for
Pepsi is different on the back of the can, while
the 7-Up unit stays the same.

PLATE 353
7-UP. The Un-Cola including a "dot" style title, using a square logo in place of the red dot.
RIGHT. CANADA DRY. Ginger Ale can with the Canada Dry logo on a red and green label.

PLATE 354
LEFT. WELCH'S GRAPE JUICE. AM/FM unit featuring their "Frozen concentrated sweetened" product.
RIGHT. MILLER HIGH LIFE. "The Champagne Of Beers" in a AM/FM version.

PLATE 355
LEFT. WATKINS BAKING POWDER. This is another unit that is somewhat rare, being limited to their dealers.
RIGHT. CAMPBELL'S TOMATO SOUP. This uses the short can mold that is also used for motor oils etc. Somewhat rare.

PLATE 356
GREEN APPLE. This is in the Tune-A-(xxx) family. It has no markings other than "Made in Hong Kong." It is 4-1/2H x 3-3/4 in diameter.

PLATE 357
GOLD APPLE. The statement "Your The Apple Of My Eye" tells me it really isn't edible - But this is a good spot for it. Made in Korea, and is 3-1/2 inches in diameter.

Courtesy BNZ

PLATE 358
BIG MAC. The sides have the famous "Golden Arches" with the name "McDONALD'S" on them. ©McDonald's Corp. (No Date). Distributed by General Electric Company (Model 2789). Made in Hong Kong, and is 4 inches square by 2-3/4H.

Courtesy BNZ

PLATE 359
FOLGER'S COFFEE. Also contains their slogan of "Mountain Grown" with the letters forming the mountain. This Hong Kong unit measures 3-3/4H x 3-1/4 diameter.

Courtesy LCB

PLATE 360
COFFEE CUP AND SAUCER. Another rather novel unit from Hong Kong, and is used with Plate 359 to get most working people going! It has a British Registration Number 985362 and measures 2-3/4H x 5-1/4 in diameter.

Courtesy WMC

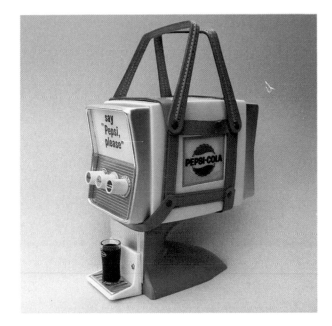

PLATE 361
FOUNTAIN DISPENSER. Say "Pepsi Please" on the front, and the Pepsi logo on the side. This unit is one of the very premium ones in this series. (See Text.) Distributed by Industrial Contacts Inc. It is made in Japan, and measures 7H x 3-1/4W x 6-1/2D.

Courtesy BNZ

PLATE 362
FRENCH FRIES. This AM version features the famous McDonald's Arches on the front. A AM/FM version is also available. Made in Hong Kong, and is 6H x 4-1/2W.

PLATE 363
GULDEN'S MUSTARD. Excellent rendering, complete with metal cap. This is the 24 Oz. size, and made in Hong Kong. It measures 5-1/2H x 4-1/4 in diameter.

Courtesy BNZ

PLATE 364
HAMBURGER/CHEESEBURGER. Another "generic" unit distributed by Amico, Windsor, and even through Sears. It is made in Hong Kong and is 2-1/2H x 6 diameter.

PLATE 365
LIPTON CUP-A-SOUP. Older unit containing "Chicken Noodle Soup." Label also states "Instant - Just Add Water." Made in Taiwan and is 5H x 3-3/4W.

Courtesy BNZ

PLATE 366
McDONALD'S COAST TO COAST. Also has "Custom Built HAMBURGERS" on the label. This is an example of a can radio used for advertising. It is marked, "Commemorative Can 1955-1985" on the back. Made in Hong Kong, and is standard can size.

Courtesy LCB

PLATE 367
MIRACLE WHIP. Also marked, "THE BREAD SPREAD" and is a product of Kraft Foods. The radio controls are via thumb wheels in the sides. Made in Hong Kong, and is 5H x 3 in diameter.

PLATE 368
NESTLE CRUNCH MILK CHOCOLATE WITH CRISPED
RICE. The side is marked, "CRUNCH IS Music to Your Mouth."
Also note "Real Chocolate" logo in the upper corner. It is 5W x
3-1/2H.

Courtesy LCB

PLATE 369
TREESWEET ORANGE. Has a cute logo with three smiling
oranges, but no other marks. Made in Hong Kong, and is 3-1/2
in diameter.

PLATE 370
OREO COOKIE. Looks like someone has already had a bite.
Although I call it an "oreo," it is only a generic type. Distributed
by Amico and ©1977. Made in Hong Kong, and is 1-3/4W x 6
diameter.

PLATE 371
PABST BLUE RIBBON BEER. This can also contains a
statement "This is the ORIGINAL Pabst Blue Ribbon Beer . . .
Selected as America's Best in 1893." Standard size.

Courtesy LCB

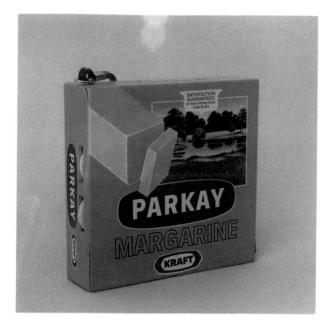

PLATE 372
PARKAY MARGARINE. This Kraft product is also marked "Satisfaction Guaranteed or your money back from Kraft." Made in Hong Kong, and is 4-1/2 square.

Courtesy LCB

PLATE 373
PINEAPPLE SPEARS IN ITS OWN JUICE. Also marked, "Great For Snacks No Sugar Added," and the Pineapple is sold by Del Monte. This radio is the same size as the "Paint" cans, 3-3/4H x 3-1/4 in diameter.

Courtesy BNZ

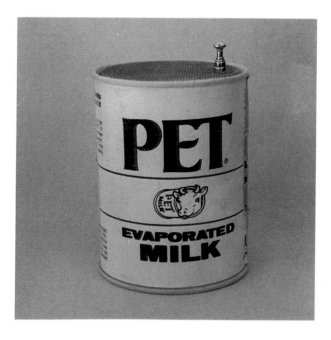

PLATE 374
PET EVAPORATED MILK. AM/FM unit with British Registration Number 1006709. Made in Hong Kong, and is 4H x 3 diameter.

Courtesy LCB

PLATE 375
RADIO CANDY. This novel unit has a British Registration Number 1009116. The controls for the unit are the flared ends of the candy. Made in Hong Kong, and is 5-3/4W x 2-3/4H.

Courtesy BNZ

PLATE 376
RED TOMATO. This unit has no other marks except "Made in Hong Kong." It is 3-1/2H x 4-1/4 in diameter.

PLATE 377
WINE CASK. This early Japanese unit is tuned by the "cork" at the top, and the volume is via the "spigot." It is 6-3/4L x 5 diameter.

PLATE 378
GRAND OLD PARR. De Luxe Scotch Whiskey. This Japanese unit has a Patent Number of 294318 (? Worn marks). The controls are via the cap. It is 5-1/2H x 2-1/2 square.

PLATE 379
SUNTORY WHISKEY. Also marked, "Finest Old Liqueur" and "Product of Japan . . . Distributed and Bottled by SUNTORY LTD." This early unit has no distributor markings and is 5H x 3-1/4W.

143

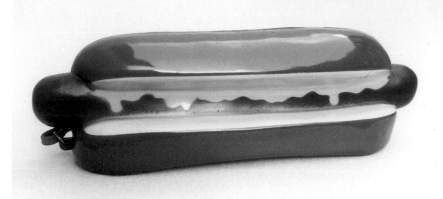

PLATE 380
HOT DOG. With mustard. This was the authors first novelty radio. As luck would have it, it is much rarer than the hamburger. The box says "Hot & Spicy sound." Distributed by Amico, and made in Hong Kong, and is 8-1/4L x 2-1/2H.

PLATE 381
ICE CREAM BAR. Looks like someone has already had a bite! Distributed by Amico, and is made in Hong Kong. It is ©1977, and is 7 inches long.

Courtesy LCB

PLATE 382
ICE CREAM CONE. This is also ©1977 by Amico. I believe it should have a base that is missing on this unit. It is made in Hong Kong, and is 7 inches long.

Courtesy LCB

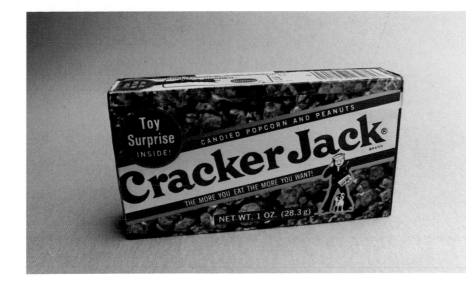

PLATE 383
CRACKER JACK. This is a horizontal version of the worlds favorite treat. See Plate 384 for another view. Shows all of the slogans: "The more you eat the more you want" and the kid's favorite: "Toy Surprise Inside." ©1968 & 1974 by Borden Inc. Made in Hong Kong, and is 5L x 2-3/4H.

Courtesy BSG

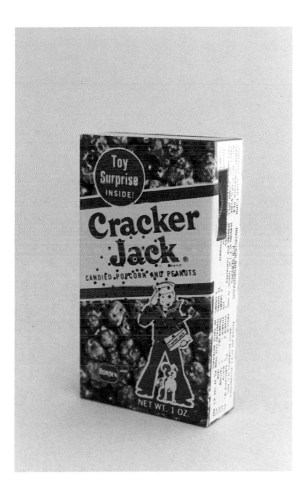

PLATE 384
CRACKER JACK. This is the other side of the unit shown in Plate 383. The format is now vertical, and it omits the famous slogan "The More You Eat - The More You Want." Same details as above.

Courtesy BSG

PLATE 385
VENDING MACHINE. "COLD DRINKS with ICE." This is a "generic" machine that was distributed by Westinghouse Co. It is a AM/FM unit, and measures 7H x 3-3/8W. (Same construction as Plate 394 - see text.)

Courtesy LCB

PLATE 386
PINCH. Scotch Whiskey in a pinch bottle.
The bottom label on the bottle says: "This
whiskey is 12 years old." Made in Hong
Kong, and is 7-1/2W x 2-3/4H.

PLATE 387
ENJOY COKE. This shirt pocket radio is made in Taiwan, and
is 3-1/2W x 2-1/2H.

PLATE 388
VENDING MACHINE. "Drink Coca-Cola" on the front and
"Have A Coke" on the side. Japanese unit marked Model TRS 618.
It is 4-1/2H x 2-3/8W.

Courtesy LCB

146

PLATE 389
VENDING MACHINE. This unit has the words "Enjoy COKE" on the front and side. A small "COKE" logo also is used above the selector buttons. It is distributed by Jack Russel Company, and is an AM/FM unit made in Hong Kong. It is 8H x 3W.

Courtesy LCB

PLATE 390
VENDING MACHINE. "Enjoy COKE" on the front and sides, along with the "Wave."©1987 by Markatron Inc. It is also ©1987 by Coca-Cola Co. Hong Kong AM/FM unit that measures 7-3/8H x 3-1/4W.

PLATE 391
VENDING MACHINE. "Enjoy Coca-Cola" on the front and "Here's The Real Thing" on the side. Made in Japan by Premium Sales Ltd., Tokyo. (Model M-P50.) It is an AM/FM unit distributed by Jack Russell Co. Inc. It is 8H x 3-1/2W.

Courtesy LCB

PLATE 392
VENDING MACHINE. This one is marked "Enjoy COKE" and "Coke is It" on the front. The side is marked "Enjoy COCA-COLA." This is a AM/FM unit distributed by Markatron Inc. (Model 2001). Made in Hong Kong, and is 7-1/2H x 3-1/4W.

PLATE 393
VENDING MACHINE. This very rough unit from 1963 originally had a decal on the front that said, "Drink Coca-Cola." The side is marked "Things go better with Coke." It uses a Westinghouse "shirt pocket" radio, housed in this vending machine case. AM/FM unit, probably made in Hong Kong. It is 7-3/4H x 3-1/2W.

PLATE 394
VENDING MACHINE. This is the inside view of the radio shown in Plate 393. You can see that the "Coke" case is simply added around an Westinghouse "shirt pocket" radio. The long red bar in the center is for the AM/FM switch.

PLATE 395
VENDING MACHINE. "DRINK COCA-COLA" and "Enjoy That Refreshing New Feeling" on the front. The side is marked "Things Go Better With Coke." Made in Japan, and distributed by Jack Russel Company Inc. It is 8H x 3-1/2W.

Courtesy LCB

PLATE 396
VENDING MACHINE. "Say PEPSI Please" on the front and also the logo "Pepsi Cola" on the case. It is distributed by Industrial Contacts Inc. and is 6-5/8H x 3W.

Courtesy LCB

PLATE 397
VENDING MACHINE. "PEPSI" with no other writing on front. The sides have only the Pepsi logo. AM/FM unit manufactured by TPI products Ltd. Distributed by Pen International Inc. with permission of Pepsico Inc. Made in Hong Kong, and is 8H x 3-1/4W.

NOTES

Music and Home Entertainment

This chapter features items that make music or entertain us in the comfort of our homes. Radios, stereos, television sets, pianos, backgammon and even dice are included in this chapter.

Items that are usually associated with outdoor recreation, such as golf, baseball etc., are shown in the chapter titled "Sports and Recreation."

While the most expensive radio appears in the "Food" section, the rarest radio in this book may well be the phonograph shown in Plate 425. It is made in Russia, and all of the markings are in cyrillic (Russian script). How this radio made its way to the United States might be an interesting story.

Another interesting radio is the "Pinball Wizard" shown in Plate 423. It is a miniature copy of an actual machine distributed by the Bally Company.

Many phonographs appear, but my personal favorite is the "Hi-Boy" style shown in Plate 424. This beautiful Japanese unit is made of plastic rather than metal, but still shows excellent detail in the horn and doors.

A very useful combination is the TV set/slide viewer shown in Plate 438. A 35MM slide can be inserted into a slot just above the screen, and when a switch on the back of the unit is turned on, the slide appears on the "TV" screen.

NOTES

PLATE 398
BACKGAMMON GAME. This Hong Kong unit is ©1978 by Amico Inc. The lid opens to reveal a storage compartment for the dice and checkers. It is 7W x 5D.

Courtesy BNZ

PLATE 399

BASS FIDDLE. This unit also doubles as a picture frame. Made in Hong Kong, and distributed by Windsor. It is 8-3/4H x 3-1/2W.

PLATE 400

GUITAR. NASHVILLE PICKER. This is a combination radio and amplified guitar. It is made in Hong Kong and distributed by Picker International. A white version is also available. It is 12-1/2H x 4-1/4W.

PLATE 401
GAMBLERS JEWEL BOX RADIO. This unit features "Dice-O-Matic" tuning. It has a wooden case with a nice felt covering depicting a crap table. Made in Japan, and distributed by Stellarwar Corp. N.Y. (Model 4342). It is 11-3/4W x 7H.

PLATE 402
DICE RADIO. This is a Japanese unit made by Sanyo (Model RP 1711). The controls are on the rear and not visible in this view. It is 3-1/4 inches square.

PLATE 403
THE GOLD RECORD. This "DiscStar FM/AM Radio" is marked model 2100. Made in Hong Kong, and is 6-3/4H x 5W.

Courtesy LCB

PLATE 404
HI-FI RACK SYSTEM. This unit has a cassette player, turntable, and tuner pictured on the front. It is an AM only unit, made in Hong Kong. The main unit is 6H x 3-1/2W and each speaker is 3-1/2H x 2-1/2W.

PLATE 405
HI-FI COMPACT SYSTEM. This system has a 8 track player, a turntable, and a AM/FM tuner pictured. It is an AM only unit. Made in Hong Kong, and the main unit is 5-1/2W x 2-1/2H. The speakers are 3-1/2H x 2-1/4W.

PLATE 406
HI-FI RECEIVER. This unit has a turntable and a AM/FM type receiver pictured. It has a U.S. Patent of 910747 and a British Registration Number of 983435. Made in Hong Kong, and the main unit is 5-1/4W x 3-3/4D. The speakers are 3-1/2H x 2-1/2W.

PLATE 407
HI-FI WITH IC POWER. This unit is another one with an 8 track player pictured, but no turntable! The radio is really an IC chip. Made in Hong Kong, and the main unit is 5-1/2W x 1-7/8H. The speakers are 3-3/4H x 2-5/8W.

PLATE 408
JANICA MINI STEREO JEWEL BOX RADIO. This unit is marked "Model SRB-12." While it has two speakers, it is AM only, and the stereo is only in your mind. Made in Hong Kong, and is 7-1/2W x 4-1/2H.

Courtesy BNZ

PLATE 409
SOUND CENTER. This state of the art unit displays about every device you could need for home entertainment: A tape deck, cassette player, television, books, and even something that could be a computer is shown. ©1986 by Nasta PowerTronic. Made in Hong Kong and is 10-3/4W x 4-3/4H.

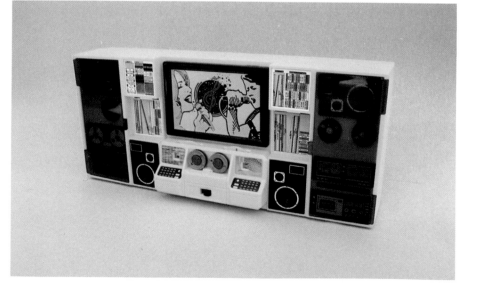

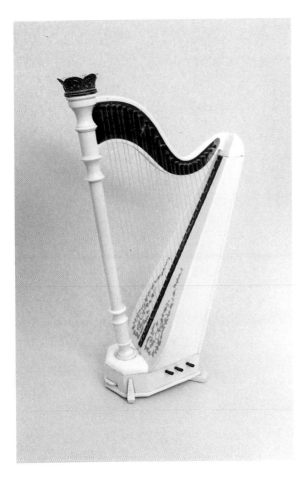

PLATE 410
HARP. This Japanese unit is distributed by Franklin and carries a U.S. Patent Number of 3170119, and also has a Japanese Patent Number of 752854. It is 11-1/2H x 4-3/4D.

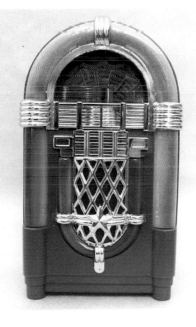

PLATE 411
JUKEBOX. This Chinese unit is distributed by Windsor as Model 380. It is an AM/FM radio that resembles the Wurlitzer 1015. It is 7H x 4W.

PLATE 412
JUKEBOX. This is also a jewelry box and a bank. It resembles a modern 45RPM type machine. Made in Japan, and is 6-1/2H x 4-1/4W.

157

PLATE 413
JUKEBOX. "Juke Voice." This Japanese unit is also a 45 RPM Machine, but possibly a later style than Plate 412. It is marked "Model JRB 32" and is 5-1/4H x 3-1/4W.

Courtesy LCB

PLATE 414
JUKEBOX. This unit resembles a Rockola model 1426. It has an AM/FM unit, and also has a cassette player behind the doors on the front. Made in Hong Kong, and is 11-1/2H x 7-1/2W.

PLATE 415
JUKEBOX. This is the official version of the Wurlitzer 1015 made in Japan and distributed by Beetland (©1986). It is AM/FM and measures 7-1/4H x 4W.

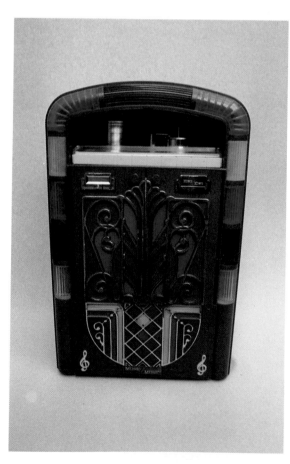

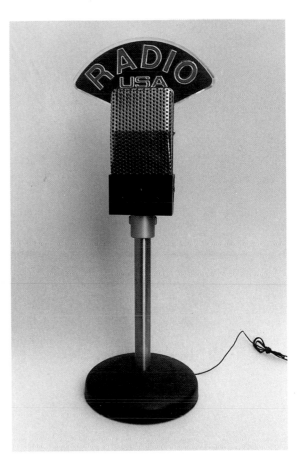

PLATE 416
MICROPHONE. "On-The-Air." This nice rendering of a 1930s style microphone is made in Taiwan and distributed by Leadworks Inc. (©1987). It is an AM/FM unit that measures 12-1/2H x 5-1/2 inches diameter at the base.

PLATE 417
MICROPHONE. "Radio USA." This microphone is also styled in the 1930s or '40s mode. It is an AM/FM unit made in China and is 16-1/4H x 5-1/2 diameter at the base.

PLATE 418
MICROPHONE. Nice octagon style mike in a plastic and metal version, made in Japan. It is marked "FBC" and is model OP-80. The controls are via the "swivel bolts" on each side of the unit. It is 8-1/2H x 4-1/4W.

Courtesy LCB

159

PLATE 419
PHONOGRAPH. GRAMY-PHONE 8. Nice example of an early phonograph with a "Morning Glory" horn. The tuning is via the record and the volume is via a thumbwheel on the side. Japanese unit that measures 5-1/2H x 4-1/2W.

PLATE 420
PHONOGRAPH. This Japanese unit is distributed by Franklin. It has a patent number of 818993. Nicely detailed unit with Brass horn, and unusual corner treatment. The base is 5 inches square and the horn is 7-1/2L x 5-1/2 inches in diameter.

Courtesy BNZ

PLATE 421
PHONOGRAPH. GRAMY-PHONE. Another version of a phonograph, but this one has a "Berliner" type horn. Same controls as Plate 419. Made in Japan, and is 7H x 6-1/2W.

Courtesy BNZ

PLATE 422
PLAYING CARDS. This side shows the King-of Hearts, while the other side is the Ace-of-Spades. Shirt Pocket style marked "Golden Sky Playing Card Radio" (No model number). No other markings, although it is probably made in Hong Kong. It is 3-1/2H x 2-1/2W.

PLATE 423
PINBALL WIZARD. This is an actual copy
of the real pinball machine. It is ©1981 by
Astra and distributed by Prestige. The
controls are via the "flippers" at the side of
the machine. Made in Hong Kong, and is 6-
1/4L x 6-1/4H.

PLATE 424
PHONOGRAPH. This "Hi-Boy" style unit with its "Nickel"
plated horn and metal trimmed doors is a plastic unit made
in Japan. Excellent detail for a plastic unit. (Note patterns
in the horn and doors.) It is 10H x 3-3/4W.

Courtesy BNZ

PLATE 425
PHONOGRAPH. This bright red unit is made in Russia and
all of the markings are in cyrillic! It is 5W x 8H.

Courtesy of LCB

161

PLATE 426
PIANO. This Grand Piano also has a 10 note keyboard to allow you to play your own tunes. It is made in Hong Kong and distributed by Newtone (Model LT-291). It is 6-1/4W x 5-1/4D.

PLATE 427
PIANO. Another version of a grand piano made by Franklin. It is marked "Copyright L. K. Rankin 1962." Made in Japan, and is 4-3/4H x 6-1/2W.

Courtesy BNZ

PLATE 428
PIANO. A wooden cased grand piano made in Japan and distributed by Lester Co. The controls are visible in this view. It is 6-1/2W x 6-1/2D.

Courtesy BG

PLATE 429
PIANO. This wooden cased grand piano is similar to the ones shown in Plates 427 & 428, but is distributed by "Tokai." Made in Japan, and measures 6-1/4W x 8D.

Courtesy WMC

PLATE 430
RADIO. This small "cathedral style" unit is made in Hong Kong and distributed by Arrow. It is 4-1/4H x 4-1/8W.

PLATE 431
RADIO. Another version of the popular "cathedral style." This one resembles a Philco model 70. Made in Hong Kong and distributed by Rhapsody. (Other versions of this radio exist.) It is 7-1/2H x 6-7/8W.

PLATE 432
RADIO. This Chinese unit is also a copy of the popular "cathedral style," and distributed by Windsor. The plate is marked "Windsor 1932 Antique Radio." It is an AM/FM unit that measures 8H x 7-1/2W.

PLATE 433
RADIO. This "tombstone" styled radio is also a cigarette box (shown open). It is made in Japan and distributed by JVC (Model JR12D). It measures 5-1/2H x 4-1/4W.

Courtesy BNZ

PLATE 434
TELEVISION WITH FISHERMAN. This
Hong Kong unit has a British registration
number of 996013. It is 6-1/2W x 4-3/4H.

PLATE 435
SLOT MACHINE. This unit, marked "Bally," is a AM/FM radio
and the handle actually works the slots. Made in Hong Kong, and
is 5-1/2H x 4W.

Courtesy LCB

PLATE 436
TAPE DECK. This reel-to-reel unit uses the reels for the controls.
It also has a visible lamp that flashes to the music. Nice detailing.
Made in Hong Kong, and is 5-3/4H x 6W.

PLATE 437
LEFT. TELEVISION WITH PICTURE OF
ELVIS. The visible knobs control the set.
Made in Hong Kong, and is 6W x 4H.
RIGHT. COLOR TELEVISION. This
Japanese unit is a pretty accurate rendering
of a color TV set. The visible knobs control
the set. It is 5-3/4W x 5H. Both sets:

Courtesy LCB

PLATE 438
TELEVISION SET. This unit is also a 35MM slide viewer, using
the screen to display the slide. It is made in Japan and distributed
by Nobility. It is 4W x 2-1/8H.

PLATE 439
WALL BOX/COUNTER SELECTOR. This very accurate
rendering is a limited edition, made by Crosley (Model CR9). The
selector buttons lift up to expose the controls. It is a AM/FM unit
measuring 11-3/4W x 13H.

Courtesy BNZ

NOTES

Sports & Recreation

This chapter features items associated with the playgrounds of America: baseball, football, basketball, and other team sports. Golf, bowling and even an outboard motor for the fishing enthusiast are also shown. Indoor recreation items are shown in the chapter titled "Music and Home Entertainment."

This chapter provides several challenges for the "series" collector. The baseball cap shown in Plate 444 is made by "Pro-Sports Marketing Inc." and uses the San Diego Padres cap resting on a small base. (In an unusual departure, the actual radio is in the cap, rather than in the base unit.) There is little doubt that one of these units exists for each professional team in the league, and possibly one for many college teams.

The baseball and basketball players

shown in Plates 445 and 446 are also part of a series, and even easier to change as the only item required is a new decal. (The box refers to licensing agreements with the various teams in the league.)

The football helmets shown in Plates 453-455 are yet another series made by Pro-Sports, and I have shown a few different ones, including a college team. Building a complete collection of these units would be an interesting challenge.

Japan is well represented in this section with the golf cart, outboard motor, baseball, and the very clever bowling ball resting on the pins — but my vote for most unusual design goes to the roller skate shown in Plate 462. This is one of the items in my collection that is often picked as being just too unusual to be a radio!

NOTES

PLATE 440
BASEBALL. This is an extra nice unit, as it has its original box. It is made in Japan by Toshiba, and is 3 inches in diameter.

Courtesy LCB

PLATE 441
DODGERS BASEBALL. This unit has no other markings, except made in Hong Kong. Possibly a "give away" at one of the home games (?). It is 3-1/4 inches in diameter.

PLATE 442
BOWLING BALL WITH PINS. This Japanese unit has a plastic bowling ball resting on 3 wooden pins. Since the pins are not attached to the ball, they are frequently missing on used units. The ball is 3 inches in diameter, and the pins are 4 inches long.

Courtesy BSG

PLATE 443
FOOTBALL WITH KICKING TEE. This unit is made in a PRI mold, so could also appear as advertising for Wilson. Since the Tee is not attached, it is frequently missing on used units. Made in Hong Kong, and the ball is 6-1/4L x 3-1/2 inches in diameter.

PLATE 444
BASEBALL CAP ON STAND. This is another unit made by ProSports Marketing Inc. This one is marked "San Diego Padres," but I'm sure there is one for each team. Made in the USA, and measures 9L x 5-5/8W.

PLATE 445
BASEBALL PLAYER. This one is the San Diego Padres, but there is one for each team in the league. The box is marked "Team Insignia is Copyright by the League." Made in Hong Kong, and distributed by Sutton Associates. (©1974). It is 9-1/2H x 5W.

PLATE 446
BASKETBALL PLAYER. Same concept and details as Plate 445, except using Basketball players. This one is the Los Angeles Lakers, but again, one for each team in the league. Not pictured is the same series showing football and hockey players.

PLATE 447
BALL RADIOS. These are AM/FM units distributed by Hyman Products Incorporated, and are ©1987. This pair is a golf and baseball. The balls all measure 4-1/2 in diameter, and are made in Hong Kong.

Courtesy BNZ.

PLATE 448
BALL RADIOS. Same details as Plate 447, but this pair is a basket and tennis ball.

Courtesy BNZ

PLATE 449
TEAM ITEMS. Here's a little different concept for the advanced collector. Try to get a Helmet and a football marked with the team logo for each team in the league.

Courtesy LCB

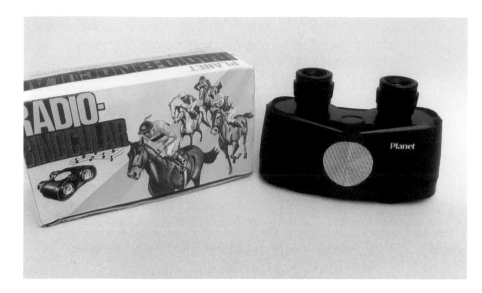

PLATE 450
BINOCULAR RADIO. A rarer unit that has a radio built into a pair of 4X binoculars. The original box shows a colorful race track scene. Distributed by Planet and made in Hong Kong. They are 5-3/4W x 3-3/4H.

PLATE 451
BALL RADIO. Same details as Plate 447, but using a soccer ball this time.

Courtesy BNZ

PLATE 452
BINOCULAR CASE. This case is made of real leather and is designed to hold a pair of 7X or 8X binoculars. An interesting variation of Plate 450, But not quite as handy to use. (Plate 450 is really a "Toy." This would house a better pair of glasses.) Made in Japan and marked "Bino-Dio." It is 8H x 7W.

Courtesy BNZ

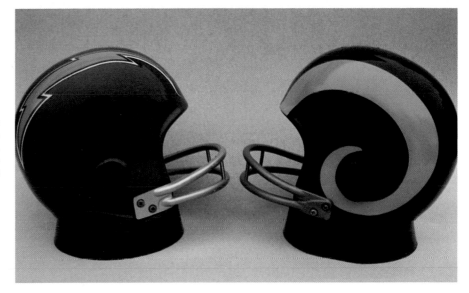

PLATE 453
FOOTBALL HELMETS. These two show the Southern California teams of the Los Angeles Rams and the San Diego Chargers. The helmets are made in the USA and distributed through ProSports Marketing Inc. They are 7W x 6-1/2H.

PLATE 454
FOOTBALL HELMETS. Same as Plate 453, but showing the NFL Vikings and the Nebraska College team.

Courtesy LCB

PLATE 455
FOOTBALL HELMETS. Same details as Plate 453, but this time using Dallas Cowboys and the L.A. Raiders. A helmet is probably made for each pro and college team in the U.S.

Courtesy BNZ

PLATE 456
FOOTBALL HELMET. This one is marked "TC" for "Tandy Corp." and is distributed through Radio Shack stores. It is made in Korea and measures 5-1/8H x 5-3/8W.

PLATE 457
GOL-PHONE. This Japanese unit has a patent number of 170406. Excellent detailing using a volume control at the end of the golf club shaft and tunes via a thumbwheel. Measures 6H x 4-1/2W.

Courtesy WMC

PLATE 458
GOLF CART WITH CLUBS. This is an early unit from Japan. The metal cart is brass plated, and the bag is plastic. Since the clubs are not attached to the bag, they are frequently missing on used units. It is 7-1/4H x 6W.

PLATE 459
HOCKEY GAMES. This is a rather compact unit showing the skates, hockey puck, and the stick. The thumbwheel controls are visible on the edge of the puck. Made in Hong Kong, and distributed by Concept 2000. It is 7H x 5-1/4W.

Courtesy WMC

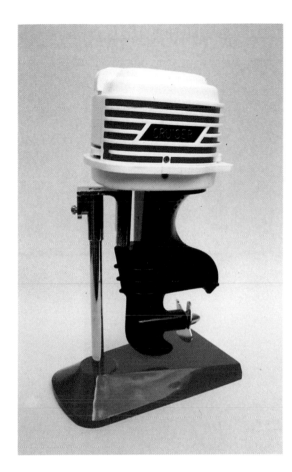

PLATE 460
OUTBOARD MOTOR. An unusual unit showing a metal stand with a plastic motor. It is marked "Cruiser" and is made in Japan. The controls are via thumbwheels that are just visible on the front and rear of the motor. (This unit is missing the "throttle" control on the front.) It is 8-3/4H x 5-1/4W.

PLATE 461
RACE TICKET. This one is from "Bay Meadows" and is for $50.00 in the 8th race to win. Other versions exist, and this radio appears to be identical to Plate 422. Probably made in Hong Kong, and is 3-1/2H x 2-1/2W.

PLATE 462
ROLLER SKATE. This unit is one that always attracts attention when people see my collection. It features bright colors and excellent detail. Made in Hong Kong, and distributed through Prime Designs. It has a U. K. Registration Number of 3442751, and measures 6H x 5W.

Today, thanks to modern technology, we have two units that would fit Mr. Gernsback's vision. One, shown in Plate 495 and labeled the "Soap and Sing," is made by Holmes. This unit, and many other similar units such as "Wet Sounds," and "Splash Dance" would fit this radio to a "T." They come in many colors, are waterproof, and even have a hole in the back to allow you to hang them on the wall while you shower. (I wonder if Holmes read this editorial?)

The unit shown in Plate 496 is somewhat different, being a combination toilet paper dispenser and radio. It also appears to be moisture-proof, comes in different colors, but lacks mobility if mounted in its proper location. It has adequate volume to fill an average bathroom with sound, even while showering.

The final tribute to "John" radios is shown in Plate 497. This unit, made in the shape of the toilet bowl, was probably not designed for a damp atmosphere, but it should survive brief periods in the bathroom. It bears the label, "The best seat in the house."

PLATE 463
BOOKCASE. This wooden cased unit also doubles as a jewelry box. It is made in Japan and distributed by Ross Co. Very nice detail work in the books. It is 7-1/4L x 4-1/8H.

PLATE 464
SET OF BOOKS. This home library has two volumes: one is "Biographical Stories of Great Composers" and the other is "Biographical Stories of Great Musicians." The lid lifts just below the titles to reveal a radio inside. Made in Hong Kong, and is 6-1/2H x 4-1/4D.

PLATE 465
BANK "Melody Coins." Bank and coin operated radio. The radio is turned on by inserting a coin into the slot. Pushing the coin through the slot turns it off and deposits the coin into the bank compartment. Wooden cased unit with metal panel. Made in Japan, and is 6-1/4W x 4H when closed.

PLATE 466
CALENDAR. Panasonic model R77. This is a perpetual type calendar with an AM radio built in. Made in Japan and measures 3-3/4H x 3-3/4W.

PLATE 467
ALARM CLOCK. This old style unit uses the "bells" for the controls. ©Holiday Fair 1968. Made in Japan, and is 4-1/4D x 2-1/2D.

PLATE 468
COUNTRY MUSIC. This rural favorite is ©1971 by H. Fishlove & Company, Chicago, Ill. The controls are via thumbwheels that are just visible on the sides. Made in Japan and is 5-1/2H x 3-3/4W.

PLATE 469
DESK SET. Pen holder and AM radio combined into a very nice appearing unit. Made in Japan, and distributed by Ross Co. (model RE711). It is 7W x 5-1/2D.

PLATE 470
GRANDFATHER CLOCK. This Japanese unit is distributed by Franklin as Model LF 210. It is 13-1/4H x 4-1/4W.

Courtesy WMC

PLATE 471
GRANDFATHER CLOCK. This unit is identical to Plate 470, but is in black plastic and has a slightly different clock.

PLATE 472
GRANDFATHER CLOCK. This large wood cased unit with visible pendulum is called the "New Englander." The clock is made in the USA, but the radio is Japanese. It is 19H x 7-1/2W.

Courtesy BNZ

PLATE 473
LAMP POST. This Japanese unit is distributed by Heritage. The plate hanging from the side can be engraved as a "presentation" award. It is 10-1/4H x 4-1/4 inches in diameter.

PLATE 474
SMOKER SOUND. Wood cased unit designed to hold cigarettes and has a compartment for a lighter. When closed it resembles a dresser or buffet. It is 9W x 5-1/2H.

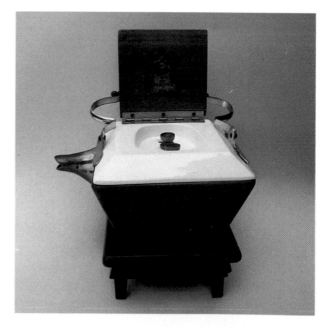

PLATE 475
TEAPOT. This is a Guild Radio product made of wood and porcelain. It is powered through the "trivet" it rests on, and measures 10-1/2H x 8W.

Courtesy BNZ

PLATE 476
JEWELBOX WITH BUTTERFLIES. This colorful unit is made in Japan and distributed by Waco. The butterflies are metal and the base is plastic. Measures 5H x 5 inches in diameter.

Courtesy BNZ

PLATE 477
NOTEBOOK. This unit also has a flashlight built into the edge. Distributed by Henica and made in Hong Kong. It is 4-1/4H x 3-1/4W.

PLATE 478
NOTEBOOK. This 8 transistor/1 diode unit is marked Model BMC 733. It appears to open from the wrong edge in this view. It is made in Hong Kong and measures 6H x 3-3/4W.

Courtesy LCB

PLATE 479
NOTEBOOK. This 3 ring binder has a very thin FM radio as part of the front cover. Distributed by Hyman Products Inc. and made in China. It is 12H x 10W.

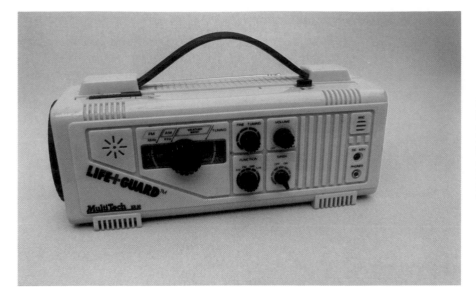

PLATE 480
LIFE GUARD. This flashlight/radio is distributed by Multi-Tech. (Model XR-20). It has AM/FM/CB and a weather channel. It is also able to be powered by turning the crank that is folded into the right hand end. Many versions of this unit, but this one has the most features of any I've seen! Made in Taiwan and is 10-1/2L x 4-1/2H.

PLATE 481
BALANCE SCALE. This metal and plastic unit is a somewhat unusual subject for a radio. It was passed by the author for 3 weeks because he didn't recognize it as a radio. (Until one of his friendly competitors bought it!) Made in Japan, and is 10-1/2L x 5-1/2H.

Courtesy BSG

PLATE 482
AMERICANA SPICE CHEST. This Japanese unit is similar to the Vacuum tube version made by Guild Radio. It is wooden cased and distributed by Audition (Model 906). It is 13-3/8H x 11-3/8W.

PLATE 483
LADIES HIGH HEELED SHOES. Two versions, one in black and one in red. This unit is somewhat unusual as it is FM only. (No AM Band.) They are made in China and ©1987 Owner Tooling. Model SH920. Also marked "Columbia Telecom. East Rockway, NY." They are 9-3/4L x 5-1/4H.

PLATE 484
PICTURE CUBE RADIO. Designed for the insertion of your family pictures. Made in Hong Kong, and is 3-3/4 square.

PLATE 485
PICTURE FRAME. This very deco looking unit is designed to house a 5 x 7 picture. It is made in Hong Kong and marked PE-397. Overall measurements are 5-1/4 x 7-1/4.

PLATE 486
CRANK STYLE TELEPHONE. This wooden cased unit has brass plated trim. The controls are via the "bells." Made in Japan, and is 11-3/4H x 9W.

PLATE 487
CAMERA. This unit is marked "De-Luxe Tour Partner" and is made in Hong Kong. Has a U.K. Registration Number of 958875 and measures 5-1/2W x 3-3/4H. (Other camera units exist.)

Courtesy WMC

PLATE 488
TELEPHONE. French Style unit decorated with cherubs on the base. (Many other versions of this style exist.) Made in Japan, and is 7H x 6W.

PLATE 489
CANDLESTICK TELEPHONE. This Korean made unit is distributed by Tandy, through the Radio Shack stores. It has outstanding control integration; using the switch hooks to turn it on, the mouthpiece is rotated for volume, and the tuning is via the dial. It is 9H x 4 inches diameter at the base.

PLATE 490
CANDLESTICK TELEPHONE. This Japanese unit is distributed by Prince. It has a small light in the mouthpiece and a cigarette lighter in the receiver. It is 10-1/2H x 5 inches diameter at the base.

PLATE 491
CANDLESTICK TELEPHONE. This Hong Kong unit comes in many colors. (Red, Black, and Green). The controls are via thumbwheels at the rear of the base.

PLATE 492
WRISTO RADIO. While it is a little large to be called a wrist watch, the principle is the same. Made in Hong Kong and distributed by Amico. It is British design number 965150 and measures 2-1/2 diameter by 1-1/4H. The strap is 10-1/2L.

PLATE 493
WRIST WATCH. This metal case unit from Japan is designed to hang on the wall with its 30 inch strap. Nice brass plating with real leather strap. The face is 4-1/2 inches in diameter.

PLATE 494
LIGHT BULB. "You Light Up My Life." Clever unit that uses screw base for volume control and thumbwheel for tuning. No distributor or origin marks on unit. It is 6-1/4L x 3 inches in diameter.

Courtesy BNZ

PLATE 495 (LEFT)
SOAP & SING. "The Shower Radio." This AM/FM unit is built in a waterproof case and has a slot in the back to allow you to hang it on the wall. It is made in Taiwan and measures 10H x 5W.

PLATE 497 (ABOVE)
LITTLE JOHN RADIO. Distributed by Church/Amico and comes in many colors. The packing case for this unit is shaped like a roll of toilet paper. British design number 965697. Made in Hong Kong, and is 4-1/4H x 4-1/4D.

PLATE 496 (LEFT)
REST ROOM RADIO. Distributed by Windsor and comes in many colors. (White, Pink, and Yellow). Its obvious function is to entertain you while you're "resting."

NOTES

Warfare, Warriors, and Weapons

This chapter features items used in man's most useless pursuit — warfare. Many years are spanned, from the "knights of olde" to a modern jet fighter.

Japanese units are well represented in this chapter, and despite the theme, the units reflect their typical high standards for detail and quality construction.

The telegraph sounder shown in Plate 500 is typical of their work, using brass-plated metal on a plastic base. The unit also doubles as a code-practice oscillator for budding amateur radio operators or Boy Scouts trying for a merit badge.

The pair of hand grenades shown in Plates 501 and 502 offers another chance to compare workmanship. The Japanese unit in Plate 501 features semi-concealed controls (under the "trigger") and plated metal. It also doubles as a cigarette lighter. The companion Hong Kong unit is plastic and lacks this attention to detail.

The "generation gap" in military fighter planes is shown in Plates 503 and 504. The bi-wing unit from Japan dates from World War I, while the Air Force F-125 represents a modern jet fighter.

Ancient warfare is represented by various knights and a castle. The knight head shown in Plate 509A is quite rare. It's a version that shows his face behind the face plate. The most common version, shown in Plate 507, has the speaker behind this plate.

NOTES

PLATE 498
CANNONS. These nicely detailed units are from Japan. The one on the left features brass plating, while the one on the right is in pewter. They have many distributors. (These are by Windsor.) They are 9L x 6W.

PLATE 499
ANCIENT CASTLE. Used by Knights of olde for refuge. This Japanese unit is made of wood, and is distributed by Heritage (Model 6801). The controls are on the roof, and it measures 10W x 5-3/4H.

PLATE 500
TELEGRAPH KEY WITH SOUNDER. This unit also doubles as a code practice oscillator. It is brass plated metal on a plastic base. Made in Japan, and is 7-1/4W x 4-1/2D.

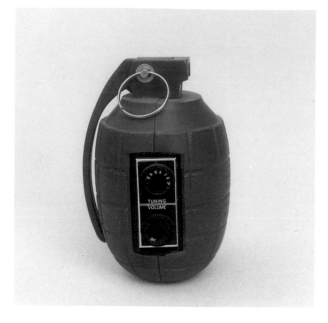

PLATE 501
HAND GRENADE. This unit has a metal base and trigger assembly. The trigger also controls a cigarette lighter. Made in Japan, and is 5H x 3-1/4 inches in diameter.

PLATE 502
HAND GRENADE. This is an all plastic unit with visible controls. Made in Hong Kong, and is 4-1/2H x 3 inches in diameter.

Courtesy LCB

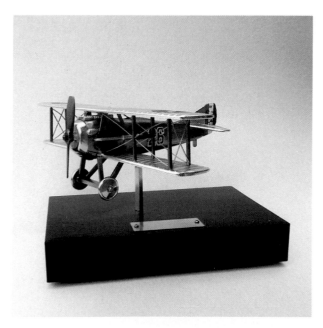

PLATE 503
BI-PLANE. This excellent rendering of a WWI fighter plane is all metal on a plastic base. Distributed by Waco and made in Japan. The base is 7-1/4W x 4-1/2D, and it is 5H overall.

Courtesy BNZ

PLATE 504
USAF F-125. This ultra modern fighter plane has no other markings. Made in Hong Kong, and is 7-3/4W x 9L.

Courtesy BNZ

PLATE 505
AUTOMATIC PISTOL. Marked "Union,"
and is a 2 transistor unit. Made in Hong
Kong, and is 5H x 7W.

Courtesy LCB

PLATE 506
G. I. JOE. This shirt pocket type radio is in camouflage paint and
bears the logo "GI Joe - A Real American Hero." Made in Hong
Kong, and distributed by Nasta Industries. It is 4-1/2H x 2-3/4W.

PLATE 507
KNIGHT HELMET. This radio has a few variations. There is
also one with a dark green base instead of the black one shown
here. Made in Japan, with pewter plated metal on a plastic base.
It measures 8-1/4H x 4-1/4W.

PLATE 508
KNIGHT ON REARING HORSE. This is a rather unusual unit, and somewhat difficult to find. It features a metal horse and knight on a plastic base that contains the radio. Made in Japan, and is 9H x 6-3/4W.

Courtesy LCB

PLATE 509
STANDING KNIGHT. Also called the "Man of LaMancha." This Japanese unit is frequently missing the sword. It also features a "presentation" plate. Distributed by Heritage and measures 11-1/2H x 6-1/4W.

PLATE 509A
KNIGHT HELMET. This is a very rare version of the unit shown in Plate 507, showing the knights face behind the face plate. Japanese unit distributed by Waco and measures 8-1/4H x 4-1/4W.

Courtesy WMC

Miscellaneous Radios

Every book that features collectibles grouped into sections seems to end up with a chapter like this one. Regardless of the number of chapters you include, a few units seem to defy classification — and most authors resort to a chapter called "Odds ands Ends, " "Can't Classify," or even "Oddballs." Although I have to follow the technique, I'll call my final chapter simply "Miscellaneous."

Many radios in this section are probably not what advanced collectors would consider true novelty sets, but are just novel — and their value to a collection is marginal. Most collectors buy these sets when they are starting to collect, and they are generally purchased to get the "count" up.

Still, a few units in this section deserve mention. The "Love" unit shown in Plate 511 uses the same technique as the "Radio" one previously shown in Plates 307 and 308. The presentation key shown in Plate 510 is the definitive example of a radio used as an award. This brass-plated unit from Japan is the only one in this book that actually has the plate engraved, and was presented to "Apeco President's Key Man."

The "CN Tower" and the "Liberty Bell" shown in Plates 513 and 514 are probably souvenirs from a trip to those locations. They are "regional" units, and their distribution may be somewhat limited — especially on the West Coast. (However, our "swap meets" and "flea markets" have lots of banks shaped like slot machines from Las Vegas and Reno, Nevada. Anyone want to swap units?)

NOTES

PLATE 510
PRESENTATION KEY. This brass plated key is from Japan, and is metal with a plastic faceplate. Notice the presentation plate on the key. It is engraved "Apeco Presidents Key Man Award." It is 14-1/2L x 6 inches in diameter.

PLATE 511
LOVE. This is in the same style as the "Radio" units. Note the visible thumbwheels. It is marked "Copyright GP8001" Made in Hong Kong, and is 7-1/4L x 3H.

Courtesy LCB

PLATE 512
CB/AM RADIO. This is a very compact little unit, featuring a CB set and a AM radio. Made in Hong Kong, and distributed by Fanon as Model 1 (©1976). It is 7W x 3-1/4H.

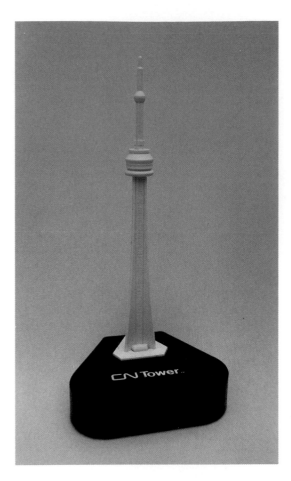

PLATE 513
CN TOWER. This is a model of the Canadian National Tower at Montreal. It is probably a souvenir from Canada. Made in Hong Kong, and is 11-3/4H x 5-3/4W.

PLATE 514
LIBERTY BELL. This unit is made in Japan and distributed by Waco. The plate mounted on the base tells the history of the bell. It is 7-1/2H x 7-1/4W.

PLATE 515
BALL AND CHAIN. This radio comes in many colors. It is made in Japan by Panasonic and is model Panapet 70. It is 4 inches in diameter.

PLATE 516
MEDALLION RADIOS. These ©1971 units feature themes of the 1960's. "Love" and "Keep It Green" shown here, but others include "Peace" and "Get Well." Distributed by Radio Shack and made in Korea. They are 4 inches in diameter.

PLATE 517
ROUND RADIO. This Concept 2000 unit has no other markings except made in Hong Kong. It is 2-1/2H x 3 diameter.

PLATE 518
G.E. SPACEBALL. This orange and clear plastic unit is very futuristic looking. It is made in Japan and distributed by General Electric Co. The base is 4-1/2 in diameter and is 5H overall.

Courtesy BNZ

201

PLATE 519
BLUE MAX. This General Electric unit, made in Japan and marked model P2760, has visible electronics thru the blue plastic cover. The box states "Visible Electronic Components are Non-functioning." Measures 4-1/2H x 3-1/2 in diameter.

PLATE 520
BIKE RADIO. This unit is sold thru Radio Shack stores, and bears the name "Archer Radar Patrol." Made in Hong Kong, and is 4-3/4W x 3H.

Courtesy WMC

PLATE 521
CHROME RADIO. This ultra modern, highly polished unit is distributed by Realistic (Radio Shack). It is marked "Lunavox Designer Series One," and made in Japan. Measures 6H x 3W.

Courtesy BNZ

PLATE 522
STOCKTICKER. This unusual unit from Japan uses brass plated metal and plastic construction. It measures 8H x 3-3/4 in diameter. (Scarce)

Courtesy WMC

PLATE 523
WORLD TIMER. This modern looking unit is actually a clock and a radio. The top turns to show the times in different parts of the world. It is marked "World Timer 747." It is 7H x 3-1/2 inches in diameter.

Courtesy LCB

PLATE 524
RED CUBE RADIO. This Japanese unit is made by Panasonic and measures 4H x 4W.

Courtesy WMC

PLATE 525
WRIST RADIO. Another of the novel units from Panasonic. This one also comes in many different colors. It is approximately 7 inches in diameter.

NOTES

Building A Collection

This chapter will try to answer the question mentioned in the introduction, namely, "Where do you find them?"

Your hunt doesn't start with a visit to "Joe's Hi-Fi and TV Parlor." Instead, a visit to Kay-Bee, Toys-R-Us, Radio Shack, or the toy department of Woolworth's will probably produce one of the current character radios. Comparison shopping among these stores will show variations in price. Enjoy this luxury now, as these units will be among the least expensive ones you will buy — and about the only time you will have a choice of *where* to buy. Spurred by this success, you will want to find another. Once you're bitten by the "collecting bug," finding a new radio to add to your collection can become an obsession.

Continuing your search can cover many areas. For the inexperienced, I will list some sources and techniques that have been successful. Sources will be ranked by prices paid.

The first thing is to tell all your friends about your new hobby. A few weeks after I started my collection I took one of my radios to work. Much to my surprise, one of my co-workers offered me a radio that he had at home gathering dust. The price was right — free. This story points out a simple but sometimes overlooked fact: friends and co-workers can be a source of radios.

Just above the gift stage are garage sales. These have become an American institution. Hundreds of items move from one household to another each weekend, proving the adage that one man's trash is another man's treasure. These will often yield radios at very inexpensive prices. These sales are patronized by dealers, so get there early, and once a radio is spotted, buy it. If you don't, it will probably turn up in a antique shop — at a much higher price.

The next best source is the "swap meet" or "flea market." These are usually a combination of many garage sales, plus a group of semi-professional dealers who sell there on a continuing basis. Use the same rules as garage sales, as the competition is even more severe. Because of increased expense, prices may be somewhat higher. This is off-set by savings in time and gasoline. On a par with these are the thrift shops run by the Goodwill, the Salvation Army, and other charitable groups.

If it has been some time between radios, the next step is to place an ad. The most successful ones will be in papers that specialize in "collector" items. "The Antique Trader" and the "Antique Radio Classified" are two of the best (listings at end of this chapter). These ads will quickly "network" you into other collectors. While most collectors would rather swap radios, especially the rarer ones, some will sell their duplicate units. The prices are higher than the previous sources, but you're dealing with people who can accurately describe the condition of the unit. No great buys, but fair ones.

The highest prices will be paid to professional dealers of "collectibles," and the worst place is at large gatherings of these dealers — such as advertising or toy shows. These dealers buy their units from all of the above sources, and then mark them up using some formula known only to them. It is among this group that the widest range of prices will occur. It is not unusual to see a common radio with a very high price. These dealers will negotiate, but they have to make a living off their products. Like most of us, they have house payments, utility bills to pay, and they probably like to eat; they are entitled a fair markup on their products. If price is not a factor, working with a dealer can produce far more radios than a hunt conducted on your own.

Many dealers have consolidated their shop into the newest rage — the "antique mall." These malls can have as few as 10, or as many as 100 or more dealers. The malls offer a wide range of items, including vintage and novelty radios.

While conducting your hunt, keep this in mind: The modern transistor radio is so compact it can be fitted into almost any device. You will quickly learn to look for the loudspeaker, as this is the hardest part of the radio to conceal. (The tuning and volume controls can be hidden in ingenious ways.) Generally speaking, the loudspeaker is at the back or on the bottom of a novelty radio and the covering grill is quite distinctive. Any item with a "carrying strap" must be investigated (many radios use these). This book should aid your search by identifying some of the units, but rule number one while visiting the above locations is to pick up any item that COULD be a radio and inspect it. Almost every collector has a tale about how he missed a "big one," and I'll relate mine:

San Diego has a very active group of radio collectors and we all haunt the same swap meets each weekend. We have a friendly rivalry, each of us is trying to find the few available radios. One of the semi-pro dealers had the Balance Scale radio for two months before one of my competitors finally took the time to examine it — and found it to be a radio! That this unit could have been passed over week after week by all of us is a testimony to the cleverness of the designers (and my own violation of rule #1).

Rule number two is to ask every seller, "Do you have any novelty radios?" You may get a blank stare in return, so you add, "A radio built in a car or a can, for example." (These are the most common ones, and the odds are the seller has seen one.) You will probably get an answer like, "Not today, but I had one a few weeks ago that looked like a _____." The blank can be anything, and I make a list of items

I have not encountered to aid my future searches.

Early in the game you will find people collecting novelty radios who are not radio collectors per se. This group, as well as the other dealers, are your competition. They're known as "crossover collectors." The term "crossover" is applied to any radio that has appeal to more than one group of collectors. Crossover is both a curse and a blessing, depending on the affluence of the crossover group. Robot radios are always priced high due to the toy and robot collectors who are conditioned to paying high prices for their collectibles. Coca-Cola is another group in this category. The blessing is from the fact that a market will always exist due to other groups competing for the radio. They also create another source for buying, selling, and trading radios.

An excellent example of a crossover item is the Champion Spark Plug radio. This is sought by radio collectors, collectors of advertising memorabilia, and would also appeal to automobile collectors. Three distinct groups bidding for this radio will always keep it priced at top dollar.

While every effort should be made to buy the best possible unit, a novelty radio, unlike many other collectibles, doesn't have to be perfect to be collectible. But always remember the golden rule when you upgrade: You must accurately describe your old unit and price it accordingly!

While I would be reluctant to pay very much for a cracked or broken radio, or one missing knobs or other cosmetic parts, I would never pass it up just because it doesn't play. Even if it doesn't play, complete units have value as a static display. Most transistor units are not played anyway as they lack fidelity. (Listening to some of the cheaper ones can almost become a mild form of torture.)

Most of the character units were designed for children. These have a very high fatality rate, the most common defect being missing battery compartment covers, and lots of use (or abuse). I've seen radios that are broken, that are missing knobs and are even melted — but they still play (a tribute to the reliability of the printed circuit board). On the other hand, I've bought radios mint-in-box that wouldn't work when a battery was first inserted.

Transistor novelty radio collecting is comparatively new as a hobby. As the hobby gains members, some will specialize in "mini-collections," perhaps only autos or advertising items. The "purist" may collect only the early metal units from Japan, while another member might specialize in the "can" radios. Most of these collections will show a mix of new and used units. Both have a place in the hobby and it is the hope of finding a new radio, or upgrading an older one, that adds joy to the hunt.

The following publications often feature novelty radios for sale or trade. Please contact the publisher at the address shown for current subscription rates.

ANTIQUE RADIO CLASSIFIED.
P. O. Box 2
Carlise, MA 01741
This is a magazine devoted exclusively to antique and vintage radio collecting. Each monthly issue features an article on a particular vintage radio, a picture section of unusual sets, and auction reports. The bulk of the magazine is classified ads (many for novelty sets).

THE ANTIQUE TRADER WEEKLY
P. O. Box 1050
Dubuque, IA 52001
A weekly newspaper covering antiques and collectibles. A wide range of antiques is covered, from glassware to coin operated machines. Has one section for "Radio, Electronic Items," but most transistor novelty sets will appear in the large "Miscellaneous" section.

NOTES

Price Guide

Price guides are useful tools for the new collector and help the older ones determine an approximate value for their collection. This guide, like all others, is just that — a guide. Believe me, no divine inspiration is represented here, but I feel the ranges shown are consistent with prices I see for novelty radios in the major publications, those offered between collectors, and a survey of several antique dealers and "swap meets" in the Southern California area. Other regions may have wide variations from these prices!

This guide, being the first one to price these units, requires some explanation of the factors I considered while trying to establish the price:

Age and Availability: Supply and demand would indicate the older units from Japan will be more expensive than the newer ones from Taiwan or Hong Kong, although some caution was used in pricing a unit high due to rarity alone. Considering the vintage units are less than 30 years old, the chance of finding units considered "Rare," stacked up in an old warehouse is very possible — and those collectors who considered only this factor could be very disappointed (the New Orleans silver dollar is an excellent example).

Desire to Own: What makes a radio desirable? Ask 10 different collectors and your likely to get 10 different answers. To the specialized collector, it's merely a matter of subject material (any Disney item, any automobile, etc.). Others may concentrate on excellent design and workmanship, while some may seek only the many "can" units. Among the factors I considered: Is it well designed and resembles the object it depicts, rather than looking like a conventional radio? The cars should look like cars and be based on units that actually were made. Unique or unusual subjects such as the coffee cup, tape measure, and the Pinball Wizard have almost universal appeal and will create desire in most collectors.

Quality of Construction: This concerns translating the chosen subject into the actual radio. Units in which the manufacturer used metal or wood construction were rated higher than those in plastic. Plated metal rated higher than painted metal, and radios made of cloth or other lesser materials rated low. AM/FM units were also given higher marks. The highest marks were assigned to those radios that used natural functions of the object for the radio controls. The Rolls Royce, using the spare tires, is a good example of control integration. The designers of the ships telegraph also did a good job by using the "speed" control for the tuning, although they had to revert to a thumbwheel for the volume control (since no other function was available, they had little choice!) Units in which the controls are not integrated, but are totally concealed rate slightly higher than those in which large knobs jut out and actually detract from the units original appearance (the Porsche auto vs. the French style phone, for example). The radios that depict the original object faithfully, or are made to scale, rate higher than those in which these factors are ignored. The Stanley tape is so lifelike that when placing an actual 25-foot measure beside the radio it is difficult to tell them apart. Other advertising items, such as the beverage and soup cans, also pass this test — but the oil cans fail. For this reason, radios that are three-dimensional rate higher than the two-dimensional ones, and the least effort seems to have been spent in the rectangular units with a gummed decal.

The above items concern the design and construction of the radio, but one more factor must be considered before we part with our hard-earned dollars for a radio: condition! Accurate descriptions of condition are very important when buying radios by mail. Probably nothing is more frustrating than buying a radio that the seller has described in "excellent" condition, yet upon opening the long-awaited package, you find a grubby, beat-up unit, possibly missing some cosmetic items too! Since most dealers in "collectibles" might not be totally familiar with each radio, I can understand the missing cosmetics, but I find it hard to believe that some of the radios I've received by mail would fit any-ones idea of "excellent." (Or if it is, I would sure hate to see what their "poor" looked like!) I have now decided I will not buy any item through the mail unless the seller agrees to refund my money if the radio fails to meet my standards. Most honest and reputable dealers, or collectors, have no problem with this if you pay the return postage: This is more than fair. If they won't agree, I automatically assume the radio is too rough for my collection.

I often wonder why this ever happens. Of all the descriptions I had to write, I found condition much easier to define than any other factor. Since this book is the first to set prices, I might as well try to establish some guidelines for rating condition. I hope most collectors can agree with the following:

The top price is for radios in "mint" condition. This means brand new, never played and in its original packing. (Often termed "mint-in-box" or "new-in-box.")

Below this is a class termed "very fine." This is a radio that is complete, playing, and has original luster with a minimum of wear. This unit probably will not have its original box, but will show careful care and handling. These radios should command about 70%-80% of the top price.

The lowest price indicated is for radios in "good" condition. This radio is one that plays, may have a few scuffs or some worn spots on the paint, and also may be missing the battery compartment cover (but no other cosmetic parts). The unit should not have any chips or cracks.

Radios that don't play; show major wear, chips, or cracks, or otherwise fail to meet the above criteria must be considered "parts" units. Values to be placed on these radios are best decided on an individual basis — and the value of the one to be fully restored is probably the key factor. (For example, I might pay dearly for the jeweled "eye" knob to make the Japanese owl complete, but probably not so much if I needed a wheel for the plastic Rolls Royce.)

Now that the guidelines and condition are defined, all that remains is to determine a range of prices for each radio. Each unit has two prices. The lower value is for radios that meet the above defined "good." The higher value is for "mint" ones. However, since the hobby is still comparatively new, these values may experience shifts as the hobby gains more members and certain items gain or lose popularity.

Plate	Item	Price	Plate	Item	Price
1.	Rolls Royce (1912)	$15-35	54.	Globe (Japanese 2 Tran.)	$20-45
2.	1912 Simplex	15-25	55.	Santa Maria	25-45
3.	1912 Model "T"	20-45	56.	Ships Telegraph	40-80
4.	Stutz Bearcat (Japanese)	25-60	57.	Sea Witch	25-45
5.	Stutz Bearcat (Hong Kong)	20-35	58.	Frigata Espanola	25-45
6.	Lincoln 1928 (Plastica)	10-18	59.	Jaguar Grill	25-40
7.	Rolls Royce (Brass)	25-75	60.	Life Preserver	15-30
8.	Rolls Royce (Pewter)	25-50	61.	747 Jet	25-45
9.	Rolls Royce (Plastic)	8-16	62.	Gas Pump 1930's (Std. Oil Co.)	12-25
10.	1931 Classic Car	8-16	63.	Gas Pump (Generic)	12-25
12.	1934 Duesenberg Model J	30-75	64.	Gas Pumps (Various)	12-25
13.	Cadillac (1963 Plastic)	20-40	65.	Matt Trakker Rig	15-25
14.	Thunderbird (1965 Plastic)	25-45	66.	"A" Team (B. A. Baracus)	15-25
15.	Mustang (1966 Fastback)	30-50	67.	Annie and Sandy	15-35
16.	Lincoln Continental (1966 Plastic)	20-50	68.	Batman	20-45
17.	Mercedes Benz Sedan	20-30	69.	Bozo The Clown	35-50
18.	Big Foot 4X4	15-35	70.	Barbie Radio System	15-30
19.	Fire Chief VW	15-35	71.	Bugs Bunny 2D	12-35
20.	Police Car VW	15-35	72.	Bugs Bunny 3D	12-25
21.	RTL 208	20-45	73.	Bugs Bunny Toothbrush Holder	15-35
22.	John Player Special	20-35	74.	Bugs Bunny Toothbrush Holder	15-35
23.	Speedway Special #3	15-35	75.	Bugs Bunny Pointing	20-40
24.	Porsche Racer #7	15-35	76.	Care Bear Cousins (Rect.)	8-18
25.	Radio Gobot	10-20	77.	Care Bears/Cousins	8-16
26.	Robochange	12-22	78.	Care Bears 2D	10-20
27.	Knight 2000	12-25	79.	Care Bears 2D	10-20
28.	Jaguar ("E" Type)	30-45	80.	Cabbage Patch Girl	12-25
29.	"Jem" Glitter Roadster	20-35	81.	Cabbage Patch Boy	12-25
30.	Ricksha	30-60	82.	Cabbage Patch Kids (Rect.)	7-15
31.	1864 Locomotive	35-50	83.	Donald Duck 2D	15-40
32.	C. P. Huntington Locomotive	20-35	84.	Donald Duck (Front Speaker)	20-45
33.	1864 Iron Horse	35-60	85.	Dukes Of Hazard	12-20
34.	1826 Locomotive	12-20	86.	Fred Flintstone	20-40
35.	The General (Locomotive)	20-35	87	The Fonz	15-30
36.	1869 Mississippi Firepumper	25-45	88.	Garfield w/Odie Charm	20-45
37.	Stage Coach (Red)	25-50	89.	Garfield "Music is my Life"	16-25
38.	Stage Coach (Brown)	25-50	90.	Goofy on Wagon	10-20
39.	Cabin Cruiser (w/Radar)	40-75	91.	Gumby and Pokey (Rect.)	12-20
40.	Cabin Cruiser	25-45	92.	Gumby (12 Inch)	25-45
41.	Mark Twain Riverboat	25-45	93.	Holly Hobbie (Start Today . . .)	18-35
42.	Belle of Louisville Riverboat	25-45	94.	Holly Hobbie w/Radio	18-35
43.	1917 Touring Car (Brass)	25-50	95.	Holly Hobbie in Rocker	20-40
44.	1917 Touring Car (Pewter)	25-40	96.	HeMan/Skeletor	10-20
45.	1917 Touring Car (Plastic)	15-30	97.	HeMan/Skeletor	10-20
46.	1908 Touring Car (On base)	15-30	98.	He Man 3D	12-25
47.	Globe (Vista)	15-35	99.	Huckleberry Hound	20-45
48.	Globe (Star Lite)	15-35	100.	Huckleberry/Yogi	20-50
49.	Globe (AM/FM)	20-40	101.	Kermit Frog	20-45
50.	Globe (Seahorse)	15-35	102.	Little Lulu	25-50
51.	Globe (Tandy)	12-25	103.	Scooby-Doo	20-40
52.	Globe (Wood)	15-30	104.	Masters Universe	12-25
53.	Globe (Old World)	15-30	105.	Marshmellow Man	12-25

Plate	Item	Price	Plate	Item	Price
106.	Mickey Mouse (Car)	$35-75	157.	Charlie Brown et al	$12-25
107.	Mickey Mouse (Nitelite)	20-50	158.	Smurf Head 2D	10-20
108.	Mickey Mouse (Guitar)	12-25	159.	Spiderman Radio System	15-25
109.	Mickey Mouse (Head)	15-35	160.	Spiderman (Round)	8-16
110.	Mickey Mouse (Chin)	20-40	161.	Spiderman 3D	25-45
111.	Mickey Mouse (Big Ears)	20-50	162.	Superman in Phone Booth	25-55
112.	Mickey Mouse (Pointing)	20-40	163.	Superman 2D	45-65
113.	Reclining Mickey	25-55	164.	Transformers	8-18
114.	Mickey Mouse (Pendant)	20-40	165.	Wrinkles	12-20
115.	Mickey Mouse (2 Trans.)	25-60	166.	SS (Rect)	8-16
116.	Mickey Mouse (Ear Tune)	8-16	167.	SS (Oven)	12-25
117.	Mickey Mouse (Armchair)	12-25	168.	Sweet Secrets	18-25
118.	Mickey Mouse Alarm Clock	20-35	170.	Tom & Jerry	25-50
119.	Mickey & Donald Duck (Nite Lite)	20-40	171.	Wuzzle (Butter Bear)	20-45
120.	Mickey Sing-A-Long (Wagon)	25-55	172.	Wuzzle (Bumble Lion)	10-20
121.	Mickey/Minnie Music City	12-25	173.	Winnie-The-Pooh	20-40
122.	Pink Panther	25-50	174.	Alligator	20-35
123.	Mork-from-Ork	15-35	175.	Adam & Eve	18-35
124.	My Little Pony	8-16	176.	Tune-A-Bear	10-20
125.	My Little Pony (Rect.)	8-16	177.	BeeGees	15-30
126.	Pacman	15-30	178.	Blabber Mouth	20-45
127.	Popeye 2D	20-45	179.	Blabber Mouse	12-25
128.	Poochie Radio System	15-30	180A.	Mr. Chatter	10-20
129.	Poochie 2D	12-20	180B.	Talking Radio	12-25
130.	Pound Puppy	8-16	180C.	Talking Radio (Blue)	10-20
131.	RAA 2D	15-30	181.	Blabber Puppy	15-25
132.	RAA (Heartshape)	12-25	182.	Tune-A-Bulldog	8-16
133.	RAA Sing-A-Long	20-40	183.	Jimmy Carter Peanut	35-50
134.	R. Ann Toothbrush	15-35	184.	Tune-A-Chick	8-16
135.	Pinocchio 2D	25-50	185.	Clown w/Balls	10-20
136.	Popples	11-22	186.	Tune-A-Cow	8-16
137.	Princess of Power	15-35	187.	Tune-A-Camel	8-16
138.	Rainbow Brite	11-22	188.	Tune-A-Hound	8-16
139.	SS Bert in Tub w/Duck	20-45	189.	Tune-A-Duck	8-16
140.	SS Lamp Post Group	15-30	190.	Tune-A-Frog	8-16
141.	SS Bert w/Alarm Clock	15-30	191.	Doll (Brunette)	30-60
142.	SS Big Bird w/Radio	15-30	192.	Doll (Blonde)	30-60
143.	SS Sing-A-Long Band	20-40	193.	Doll (9 Inch)	30-60
144.	SS Big Bird Head 2D	8-16	194.	Tune-A-Elephant	8-16
145.	SS Big Bird on Nest	15-25	195.	Happy Face Radio	5-15
146.	SS Cookie Monster/Oven	15-30	196A.	Don't Touch Dial	15-30
147.	SS Oscar The Grouch	10-20	196B.	Infinity	15-30
148.	SS Bert & Ernie Center Stage	10-25	197.	Lady Bug (Phono)	12-25
149.	SS Bert & Ernie Pals	10-20	199.	Lady Bug (Wings Move)	12-25
150.	Six Million $ Man Backpack	12-20	200.	Michael Jackson (Rect)	10-20
151.	Snoopy on Doghouse	15-40	201.	John Lennon	20-35
152.	Snoopy (Outline)	8-16	202.	Hello Kitty	8-16
153.	Snoopy Pointing	20-40	203.	Tune-A-Leo	8-16
154.	Snoopy on Rocket	25-50	204.	Lady in Hoop Skirt	100-200
155.	Shirt Tales	15-25	205.	Love Is..For Us	12-25
156A.	Smurfette 2D	10-20	206.	Lucky Horse Shoe	12-25
156B.	Smurf 2D	8-16	207.	Male Chauvinist Pig	20-40

Plate	Item	Price	Plate	Item	Price
208.	Boy w/Candle	$15-35	262.	Bon-Ami Cleanser	$20-35
209.	Boy w/Suitcase	15-35	263.	Carrier High Efficency	20-35
210.	Boy w/Waterbuckets	15-35	264A.	Big A Battery	12-25
211.	Money Talks	20-45	264B.	Maintenance Free Battery	12-25
212.	Boy Playing Instrument	15-35	265A.	Atlas Battery	12-25
213.	Tune-A-Sheep	8-16	265B.	Mobil Premier Battery	12-25
214.	Tune-A-Monkey	8-16	266A.	Texaco Super Chief	12-25
215.	Monkey Head 3D	10-25	266B.	Co-op Forget-it	12-25
216.	Mouse on Base	8-16	267.	Delco Freedom Battery	12-30
217.	Tune-A-Pig (Bank)	10-20	268.	Radio Shack "D" Cell	15-25
218.	My Country Kid	10-20	269.	Ray-O-Vac "D" Cell	15-35
219.	Owl (Japanese/Metal)	30-60	270.	Carter Peanut Can	20-35
220.	Owl (Plastic)	10-20	271A.	Souptime Soup	10-20
221.	Elvis Presley (Thunderbird)	25-50	271B.	Champion Plugs	10-20
222.	Panda Bear 3D	10-20	272.	Charlie Tuna (BiCycle)	35-50
223.	Panda Bears 2D	20-35	273.	Charlie Tuna	40-60
224.	Elvis Presley (White)	25-50	274A.	Marlboro (Soft Pack)	20-35
225.	Rabbit 2D	15-25	274B.	Marlboro (Flip Top)	20-40
226.	Tune-A-Rabbit	8-16	275.	Kent Cigarettes (Large Box)	60-125
227.	Statue of Liberty	20-40	276.	Kent (Flip Top)	20-40
228.	Radio Shirt	20-35	277.	Dermoplast	15-30
229.	Snorks	12-25	278.	Eveready Classic	12-25
230.	Turtle (Plastic)	25-50	279.	GE Best Buy	10-20
231.	Spirit of '76	12-25	280.	Faultless Starch	15-35
232.	Tune-A-Tiger	8-16	281.	Texaco Filter	10-25
233.	John Wayne	25-50	282.	Getty Gasoline Pump	15-25
234.	Tune-A-Whale	8-16	283A.	Amoco Gas Pump	15-25
235.	Armillary Sphere	20-40	283B.	Sinclair Gas Pump	20-35
236.	Creature I	20-40	283C.	Sunoco Gas Pump	15-25
237.	Cylon Warrior	15-25	284.	Hamburger Helper Hand	25-45
238.	RadioBot	8-16	285.	Hoelon Herbicide	12-25
239.	Times Sputnik	15-25	286A.	Texaco Gasohol	20-35
240.	Moonship	15-25	286B.	Texaco Fire Chief	15-25
241.	Transformers (Freedom Fighter)	8-16	287.	KOSI Lite	12-25
242.	Transformers (Face)	12-25	288.	Opticurl Acid Wave	20-35
243.	Starroid IM1	30-60	289A.	Texaco Havoline Oil	10-20
244.	Starroid IR12	30-60	289B.	Mobile Delvac Oil	10-20
245.	Robark	30-60	290A.	Havoline Super Premium	10-20
246.	Dark Invader	25-50	290B.	Shell Super X	10-20
247.	Aquastar	25-50	291A.	Mobiloil Super	10-20
248.	Star Explorer	25-50	291B.	Chevron Custom (Fuel Saving)	10-20
249.	Space Shuttle (Columbia)	12-25	292A.	Mobil 1	10-20
250.	Robotic Radio	10-20	292B.	Chevron Custom Oil	10-20
251.	Robo-AM1	10-20	293A.	STP (Racers Edge)	10-20
252.	Starroid IR4U	30-60	293B.	STP (Secret Formula)	10-20
253.	Robot w/Clock	25-50	294.	Punchy	20-40
254.	RoboChange	12-22	295.	Union 76 Ball	12-25
255.	Radio GoBot	10-20	296.	Shell Oil Logo	75-100
256.	Mr. D. J.	40-60	297.	Chain Saw Oil	15-25
257.	Space Capsule	10-20	298.	NEWS Vendor	15-30
258.	Flying Saucer	8-15	299A.	Boysen Paint Can	15-30
259.	Power Box Robot	10-20	299B.	Diamond Vogel Paint	15-30
260.	Borax Soap	20-40	300A.	Cook Paint Can	15-30

Plate	Item	Price	Plate	Item	Price
300B.	Pratt & Lambert Paint	$15-30	345B.	Mello Yello	$10-20
301.	Planters Peanuts	30-45	346A.	Coca-Cola (Glossy Wave)	8-16
302.	Pillsbury Doughboy	15-25	346B	Royal Crown "RC"	12-25
303.	Raid Clock/radio	40-80	347A.	Coca-Cola (Wave 12oz.)	10-20
304.	Rust-Oleum Spray Paint	15-30	347B.	Dads Root Beer	10-20
305.	California Raisin Man	15-35	348A.	Mug Root Beer (In Foam)	12-25
306.	Safe	15-30	348B.	Mug Root Beer (On Mug)	10-20
307.	Radio (Red)	15-35	349A.	Barrelhead Root Beer	10-20
308.	Radio (Avon)	15-35	349B.	Pepsi (Modern Logo)	10-20
309.	Radio Shack (Logo)	20-35	350A.	Pepsi Light	10-20
310A.	KMPC 710 AM	10-20	350B.	Coors	8-16
310B.	Las Vegas Stars (KORK AM)	12-25	351A.	Pepsi (New Style AM/FM)	12-25
311.	Safeguard Soap	15-25	351B.	7-Up (Red Dot AM/FM)	12-25
312.	Schiltz Football	10-25	353A.	7-Up (Dots)	10-20
313.	Spark Plug (Small AM/FM)	45-60	353B.	Canada Dry	12-25
314.	Spark Plug (Large AM)	60-125	354A.	Welch's Grape (AM/FM)	12-25
315.	Spark Plug (Large AM/FM/AC)	70-150	354B.	Miller High Life (AM/FM)	12-25
316.	Stanley Tape Measure	20-45	355A.	Watkins Baking Powder	20-40
317.	Tony The Tiger	20-35	355B.	Campbell's Tomato Soup	10-25
318.	Steel Belted Tire	15-30	356.	Green Apple	10-20
319.	Winston "Winner" Tire	20-35	357.	Gold Apple	15-25
320.	Little Sprout	20-45	358.	Big Mac Box	12-25
321.	Woolite Bottle	20-45	359.	Folger's Coffee	15-35
322.	Zany perfume	20-45	360.	Coffee Cup/Saucer	20-40
323.	Ballentines Scotch	50-75	361.	Pepsi Fountain Dispenser	175-400
324.	Budweiser Beer Bottle	15-25	362.	French Fries (McDonald's)	15-30
325.	Coke Bottle	12-25	363.	Gulden's Mustard Jar	25-50
326.	Camus Grand VSOP	25-50	364.	Hamburger/Cheeseburger	12-25
327.	Hershey's Syrup	20-40	365.	Lipton Cup/Soup	15-30
328.	Heinz Tomato Ketchup	30-50	366.	McDonald's Coast/Coast	12-25
329.	Fleishmans Gin	35-60	367.	Miracle Whip	15-30
330.	Malibu Rum Bottle	20-35	368.	Nestle Crunch	15-35
331.	Pepsi Bottle	10-20	369.	Treesweet Orange	12-25
332.	Piper Brut Champagne	20-50	370.	Oreo Cookie	15-30
333.	Teacher's Highland Cream	20-50	371.	Pabst Blue Ribbon	8-16
334.	Yago Sant Gria	25-45	372.	Parkay Margarine	15-35
335.	Coca-Cola Billboard	15-30	373.	Del Monte Pineapple	20-35
336.	Pepsi Cola Billboard	15-30	374.	Pet Evaporated Milk	15-35
338A.	Budweiser (King/Beers)	8-16	375.	Radio Candy	20-35
338B.	Seagram's Cooler	10-20	376.	Red Tomato	12-25
339A.	Carnation Hot Cocoa	10-20	377.	Wine Cask	30-50
339B.	Seven-Up (Vertical)	12-25	378.	Grand Old Parr	30-45
340B.	Sunkist Orange	10-20	379.	Suntory Whiskey	40-60
341A.	Tecate Beer	12-25	380.	Hot Dog	20-35
341B.	Schlitz Beer	8-16	381.	Ice Cream Bar	20-35
342A.	Coors (AM/FM)	12-25	382.	Ice Cream Cone	20-35
342B.	Budweiser (King/Beers AM/FM)	12-25	383.	Cracker Jack Box	20-35
343A.	Coors Light (Aluminum)	10-20	385.	Cold Drinks w/Ice	35-50
343B.	Hamm's Beer	10-20	386.	Pinch	20-35
344A.	Coca-Cola (Olympic)	12-25	387.	Enjoy Coke	8-16
344B.	Fresca	10-20	388.	Drink Coca-Cola . . .	75-125
345A.	Coke (Japanese)	20-45	389.	Enjoy Coca-Cola . . .	60-120

Plate	Item	Price	Plate	Item	Price
390.	Enjoy "Coke" w/Wave	$25-50	442.	Bowling Ball w/Pins	$15-25
391.	Enjoy Coca-Cola (Bottles/side)	60-125	443.	Football w/Kicking Tee	12-25
392.	Enjoy "Coke" (Coke is It)	35-50	444.	Baseball Cap	15-35
393.	Drink Coca-Cola . . . (1963)	75-150	445.	Baseball Player	12-25
395.	Things go Better w/Coke . . .	75-150	446.	Basketball Player	12-25
396.	Say Pepsi Please . . .	100-150	447A.	Golfball	12-25
397.	Pepsi Vending	40-60	447B.	Baseball	12-25
398.	Backgammon Game	20-40	448A.	Basketball	12-25
399.	Bass Fiddle	20-45	448B.	Tennis Ball	12-25
400.	Nashville Picker	20-35	449A.	Bronco Helmet	12-25
401.	Dice-O-Matic Table	25-40	449B.	Bronco Football w/Base	10-20
402.	Dice Radio (Cube)	10-20	450.	Binoculars	20-40
403.	Gold Record (AM/FM)	15-25	451.	Soccer Ball	12-25
404.	Hi-Fi Rack System	10-20	452.	Binocular Case	20-40
405.	Hi-Fi Compact System	10-20	453A.	Charger Helmet	12-25
406.	Hi-Fi Receiver	10-20	453B.	Rams Helmet	12-25
407.	Hi-Fi w/Ic power	10-20	454A.	Nebraska Helmet	12-25
408.	Janica Mini-Stereo	12-25	454B.	Vikings Helmet	12-25
409.	Sound Center	12-25	455A.	Cowboys Helmet	12-25
410.	Harp (Japanese)	25-50	455B.	Raiders Helmet	12-25
411.	Jukebox (Windsor)	15-30	456.	Team Tandy Helmet	10-20
412.	Jukebox (45 Rpm Japanese)	25-55	457.	Golf-Phone	25-55
413.	Jukebox "Juke Voice" (Japanese)	25-55	458.	Golf Cart w/Clubs	20-45
414.	Jukebox (Rockola)	30-60	459.	Hockey Games	15-30
415.	Jukebox (1015 Wurlitzer)	40-60	460.	Outboard Motor	35-60
416.	"On-The-Air" Mike	25-75	461.	Race Ticket	10-20
417.	"Radio USA" Mike	35-70	462.	Roller Skate	15-30
418.	Mike (FBC Japanese)	25-50	463.	Bookcase w/Books	15-25
419.	Gramy-Phone 8	20-40	464.	Set of Books	10-20
420.	Phonograph (Brass Horn)	35-60	465.	Melody Coins Bank	10-20
421.	Phonograph (Berliner Style)	20-40	466.	Panasonic Calendar	10-20
422.	Playing Cards	10-20	467.	Alarm Clock	15-25
423.	Pinball Wizard	30-45	468.	Country Music	12-25
424.	Phonograph (Hi-Boy Style)	25-50	469.	Desk Set w/Pen	10-20
425.	Phonograph (Russian)	25-(?)	470.	Grandfather Clock (White)	25-45
426.	Piano (w/10 note keyboard)	12-25	471.	Grandfather Clock (Black)	25-45
427.	Piano (Grand/Franklin)	65-100	472.	Grandfather Clock (Wood)	40-75
428.	Piano (Grand/Lester)	65-100	473.	Lamp Post w/Light	15-35
429.	Piano (Grand/Tokai)	65-100	474.	Smoker Sound	10-25
430.	Radio (Small Cathedral)	8-16	475.	Tea Pot (Guild)	60-120
431.	Radio (Cathedral Model 70)	10-20	476.	Jewel Box w/Butterflies	20-35
432.	Radio (Cathedral/Windsor AM/FM)	15-25	477.	Notebook (w/Flashlite Edge)	10-20
433.	Radio (Tombstone Style)	20-40	478.	Notebook (8 Transistor)	10-20
434.	Television Set (w/Fisherman)	15-25	479.	Notebook (3 Ring)	15-30
435.	Slot Machine	25-55	480.	Life Guard (Flashlite)	12-25
436.	Tape Deck (Reel/Reel)	10-25	481.	Balance Scale	35-60
437A.	Television Set (w/Elvis)	15-25	482.	Spice Chest	15-30
437B.	Television Set (Color Console)	15-25	483.	High Heel Shoes	15-30
438.	Television Set (w/Slide Viewer)	15-35	484.	Picture Cube	10-20
439.	Wall Box/Counter Selector	80-95	485.	Picture Frame (Deco)	10-20
440.	Baseball (Toshiba)	15-35	486.	Telephone (Wall Style)	15-30
441.	Baseball (Dodgers)	10-20	487.	Camera (35MM type)	15-35

Plate	Item	Price	Plate	Item	Price
488.	Telephone (french w/Cherubs)	$15-30	507.	Knight Helmet	$20-45
489.	Telephone (Candlestick/Tandy)	15-30	508.	Knight On Horse	25-55
490.	Telephone (Candlestick w/Lighter)	15-30	509.	Standing Knight	20-45
491.	Telephone (Candlestick/Hong Kong)	10-20	509A.	Knight Helmet w/Face	45-75
492.	Wristo Radio	10-20	510.	Presentation Key	30-60
493.	Wristwatch (30″ Strap)	25-55	511.	LOVE	12-25
494.	Lightbulb	20-40	512.	CB Radio	8-16
495.	Soap & Sing	10-20	513.	CN Tower	20-45
496.	Restroom Radio	10-20	514.	Liberty Bell	25-45
497.	Little John Radio	15-35	515.	Ball & Chain	8-16
498A.	Cannon (Brass/Japan)	20-40	516.	Medallion Radios	10-20
498B.	Cannon (Pewter/Japan)	15-35	517.	Round Radio	8-16
499.	Ancient Castle	15-35	518.	GE Spaceball Radio	10-20
500.	Telegraph Key w/Sounder	75-125	519.	Blue Max	10-20
501.	Hand Grenade w/Lighter	20-45	520.	Bike Radio	8-16
502.	Hand Grenade (Plastic)	12-25	521.	Chrome Radio	10-20
503.	Bi-Plane (On Base)	40-70	522.	Stock Ticker	50-90
504.	USAF F-125	15-30	523.	World Timer	10-20
505.	Automatic Pistol	60-100	524.	Red Cube Radio	10-20
506.	GI Joe	8-16	525.	Wrist Radio	8-16

Bibliography

Alth, Max, **COLLECTING OLD RADIOS AND CRYSTAL SETS.** Wallace-Homestead Books, (Des Moines, Ia.) 1977. Covers early battery and crystal sets from 1920's thru 1930's. Brief history of radio and selected manufactures.

Collins, Philip, **RADIOS THE GOLDEN AGE.** Chronicle Books, (San Francisco, Ca.) 1987. Excellent photographs of over 100 art-deco and novelty vacuum tube sets from the 1930's to the 1950's.

Gernsback, Hugo, **"Radio In The Bathroom."** Editorial outlining future of radio in the bathroom. Radio-Craft magazine, October, 1934 issue.

Johnson, David and Betty, **ANTIQUE RADIOS RESTORATION AND PRICE GUIDE.** Wallace-Homestead Books, (Des Moines, Ia.) 1982. Many pictures of vintage sets, but known for the excellent section on repair of the vacuum tube set.

Longest, David, **CHARACTER TOYS AND COLLECTIBLES.** (Series I & II). Collector Books, (Puducah, Ky.) 1984 & 1987. Covers media related toys and collectibles. (Popeye, Mickey Mouse Etc.)

Marrs, Wanda and Texe, **THE GREAT ROBOT BOOK.** Little Simon, (New York, N.Y.) 1985. History of robots from movie monsters to modern workplace helper.

McMahon, Morgan, **VINTAGE RADIO 1887-1929.** Vintage Radio, (Palos Verdes, Ca.) 1973. This book is almost the bible for vintage radio collectors. Has over 1000 pictures of early sets and a very definitive history of the early days of radio.

McMahon, Morgan, **A FLICK OF THE SWITCH 1930-1950.** Vintage Radio, (Palos Verdes, Ca.) 1975. This book continues the saga of collectible radios and includes early Television. Over 1000 pictures.

Stokes, John, **70 YEARS OF RADIO TUBES AND VALVES.** Vestal Press Ltd. (Vestal, N.Y.) 1982. The best book on the history of the radio tube. Practically dates every tube developed. Many pictures.

Wolff, Michael F. (Editor) **"THE SECRET SIX MONTH PROJECT."** Magazine article on the development of the Regency TR-1 radio. IEEE Spectrum, December, 1985 issue.

Unknown Author, **"A NEW SHIRT POCKET RADIO."** Magazine article introducing the Regency TR-1 radio. Radio & Television News, January, 1955 issue.